STRIKE AND STRIKE AGAIN

455 SQUADRON RAAF 1944-45

IAN GORDON

ECHO BOOKS

First published in 1995 by Banner Books.

This second edition has been published by Barrallier Books Pty Ltd, trading as Echo Books, with the kind permission of Banner Books.

Barrallier Books registered office: 35-37 Gordon Avenue, West Geelong, Victoria 3220, Australia.

www.echobooks.com.au

National Library of Australia Cataloguing-in-Publication entry: (paperback)

Creator: Gordon, Ian, 1952- author.

Title: Strike and strike again : 455 Squadron RAAF 1944-45 / Ian Gordon ; Catherine Gordon (cartography and charts).

Edition: Second edition.

ISBN: 9780994355850 (paperback)

Notes: Includes index.

Subjects: Australia. Royal Australian Air Force. Squadron, 455.

World War, 1939-1945--Regimental histories--Australia.

World War, 1939-1945--Europe--Aerial operations, Australian.

Other Creators/Contributors: Gordon, Catherine G.

Dewey Number: 940.544994

Cover design and book preparation by Tim Millhouse

About the Author

Ian Gordon was born in Perth, Western Australia, and he served for nearly 40 years in the Australian Army. His interest in 455 Squadron began with his father, Ivor, who served with the Squadron as a Beaufighter navigator over much of the period of this story. Ian retired from the Australian Army in 2009 and now owns Barrallier Books Pty Ltd, also trading as Echo Books, www.echobooks.com.au.

Contents

Maps and Illustrations

The maps and sketches have been derived using British Official War Office Maps, commercial maps and sketches from the Australian Commonwealth Archives. They were all drawn by my sister Cathy Gordon, whose considerable talent has enhanced this presentation.

There is some variation in the spelling of certain place names across various references. For consistency, spelling that was in common usage in Coastal Command during the Second World War has been used.

Foreword

Wing Commander J.N.Davenport
AC, DSO, DFC and Bar, GM, MID

IN its beginnings, 455 Squadron flew the Handley Page Hampden aircraft in Bomber Command from bases in England. Then the Squadron converted the Hampdens to carry torpedoes for daylight operations with Coastal Command. Later, 455 Squadron converted to Bristol Beaufighters, equipped with four cannons and eight rockets in daylight low level anti-shipping strikes.

Ian Gordon has written an outstanding history of the Beaufighter operations and activities of this Australian Squadron. He has brought home the excitement, the dry throats, the hardship, the courage, the service and the dedication of the pilots, the navigators and the ground crew. Fear is rarely mentioned but it's there. The comradeship comes through.

It is a story of the remarkable skill and dedication of ground crew who achieved the highest level of serviceability in conditions which were frequently appalling. They worked in the open, in freezing conditions with snow and sleet, particularly in northern Scotland where winds at times were horrific. They would see their beloved Beaufighter return from an operation, shot up, perhaps severely damaged, even crash landing. The importance of returning aircraft to the 'Available List' despite the weather was well understood by these remarkable men – they achieved miracles. Australian ground crew were a race apart.

I always had great admiration for navigators, they were often the unsung heroes. In a Beaufighter he was firstly navigator, but also wireless operator and rear gunner. He was also required to operate the large hand held camera and act as controller and eyes of the pilot to protect their aircraft from fighter attack from the rear. However they had little chance of knowing what was happening up front. I owe a great deal to Flying Officer Ralph Jones DFC and Bar, RAF, my navigator in Beaufighters. He was a skilled navigator and the tougher the going, the better his navigator logs seemed to be. He always seemed calm and we had a great understanding. My navigator in Hampdens both in Bomber Command and Coastal Command, Alan Bowman saved me and my aircraft many times, but that is another story.

The Beaufighter with its cannons and rockets was a formidable strike weapon. Its rugged construction comes through in this story as so many crash landed and crews survived.

The men of the Squadron were young and enthusiastic and able. The pilots were of a very high calibre, aggressive and daring - they contributed their all and the development of the ever changing tactics belonged to them - they all had an input. I was fortunate to have served with some wonderful people.

What is important is that Ian Gordon AM, a distinguished Colonel in the Australian Defence Force, has told the story in a fascinating way using the individual accounts and reports of many of those directly involved as well as official records. His father, Pilot Officer Ivor Gordon served as navigator and joined 455 Squadron in mid 1944. His considerable involvement is treated with restraint by his son.

I am privileged to write this Foreword to a part of history.

Jack Davenport
Mosman April 1994

Preface

A USTRALIAN airmen made a dramatic contribution to the Allied war effort during the Second World War. At some time or other, Australians served with practically every combatant air squadron in the Royal Air Force and there were seventeen Royal Australian Air Force squadrons operating in Europe by the end of the war.

This is the story of one of those Australian squadrons, one of the most successful. 455 Squadron, Royal Australian Air Force operated first with Bomber Command and then with Coastal Command. In the last eighteen months of the War the Squadron flew Bristol Beaufighters in attacks against German shipping along the Dutch and Norwegian coasts. Although the Squadron usually had no more than twenty aircraft, it sank many thousands of tons of enemy shipping. It was difficult, gruelling and costly work. That last eighteen months cost the lives of forty-six 455 Squadron aircrew. This story concentrates on the Squadron's Beaufighter period. The men were called 'The Viking Boys' and the Squadron motto was 'Strike and Strike Again'.

Much of what is extraordinary in this story lies in its circumstances. Those circumstances brought together many young Australian airmen for years of training in Australia, Canada and Britain. Then they made their way to this Squadron, to fly British strike aircraft over Dutch and Norwegian coasts. No less remarkable are the tides of measure and counter-measure that can be followed as the strike squadrons sought success, and the German shipping made their defence. The skill and courage of the airmen in their work will always command recognition.

I struggled for some years to find the way of telling the story. It finally came together through the voices and recollections of the men from the Squadron. I have done my best to represent the 455 Squadron story as accurately as official records and memories will allow. Sometimes there were contradictions, and where there was uncertainty I have given the personal recollections primacy.

In completing this story I have had an enormous amount of help from the men of 455 Squadron and their families. They have resurrected log books, photographs, diaries and memories. They have corresponded, encouraged, and corrected numerous drafts. I thank Jack Davenport from Mosman, Grant Lindeman from Armidale, Neil Smith from Naracoorte, David Whishaw from Carrick, Peter Ilbery from Farrer, John Ayliffe from Kings Park, Jack Tucker from Balwyn, Bill Herbert from Flynn, the late Noel Turner from Bateau Bay, Jack Cox from Gosford, Allan Ibbotson from Rockingham, Scott Milson (Colin's son) from Bondi, Madge Watson (Doc's wife) from Carey Bay, Jack McKnight from Pennant Hills, Alan Carter from McCrae, Jock

Berry from Phillip Island, Lloyd and Thelma Wiggins from Aldgate, Audrey Sides (Fred's sister) from Black Rock, Wally Kimpton from Mount Lawley, Ian Murray from Noosa, Vic Pearson from St Lucia, Bert Iggulden from Birmingham, Helen Clifford (Lloyd's wife) from Leabrook, Dick and Neil McColl (Bob's sons), Katie Ward from Norfolk, the late Harold Spink, Jack MacDonald from East Melbourne, Forbes Macintyre from Melbourne, Ralph Jones from Croydon (UK), Bernard Collaery (son of Ted) from Canberra, Bill Bremner from Red Hill, Leonore Macbeth (Bob's wife) from Gordon, New South Wales, Bill Waldock from Karrinyup, Ray Dunn from Kyneton, Reg Andrew from Florida Gardens, Isabel Harrison from Buckie in Banffshire, Scotland, and my own father Ivor Gordon from Swanbourne. My thanks go also to Lex McAulay from Banner Books for his help and encouragement. I am very grateful for the great talent of my sister Cathy Gordon who drew all the maps and diagrams, to Ronda for typing, to Neil and Margaret for their editorial help and to Ula, my wife, who persevered through the research and the many revisions.

A word as to pronunciation. The Squadron was Four five five Squadron in keeping with the traditional manner of speaking of the Royal Australian Air Force. A reference to four hundred and fifty-five or four fifty-five brands the speaker as one who knew not the ways and customs of the operational stations.

Flight Lieutenant John Lawson, Unit Adjutant and the author of 'The Story of 455 (RAAF) Squadron'

Ian Gordon
Melbourne 1994

1

455 Squadron
The Hampden Years

It is no wonder that at this hour of...real peril, Great Britain should have turned to...the Dominions, and to us perhaps not least of all.

Australian Prime Minister Robert Menzies, Broadcast in Melbourne, 11 October 1939.

It was a most frightening and hair raising operation. We had to fly at low level and the strong turbulence made it dangerous. We sighted nothing and by the time we returned most of the crews were ill.

Flight Lieutenant Jack Davenport, 455 Squadron pilot, on 455 Squadron's expedition to Russia in September 1942.

The torpedo aircraft went straight for the merchant ships in the convoy while their escorts dealt with the enemy's escorting ships with bombs, cannon–fire and rockets.

RAF Coastal Command Review describing strike wing tactics.

A T the start of the Second World War, imperial sentiment in Australia was very strong. Britain declared war on Germany on 3 September 1939 and Australia followed almost immediately. The next day, Monday 4 September, the Editor of *The West Australian* newspaper expressed his thoughts,

The Dominions have grown into nations, proud of their nationhood... To fall under Nazi influence would mean the destruction of its [the British Empire's] inner life of ordered freedom...No nation has more to gain than Australia by the ridding of the world from the fever of aggression that has been poisoning its life...

By 20 September, the Australian Government had decided to send an expeditionary air force of six squadrons overseas before the end of that year (the formation of the 6th Division of the Second Australian Imperial Force also had just been announced, on 15 September). However, this plan for an expeditionary air force was complicated by the fact that it would absorb many of the men needed to train new recruits. On top of that, the aircraft held by the Royal Australian Air Force, its Demons, Seagulls and Ansons, were obsolete.

At the same time, the Royal Air Force was in a desperate situation. One of the main weaknesses of the Allies was in air strength, and British planners were expecting that the war would 'consume' fifteen to twenty thousand pilots each year. Britain had neither the population nor the training capacity to absorb this and it seemed that only a massive contribution of aircrew by the dominions could turn the Royal Air Force into a viable fighting service.[1]

The use of dominion manpower in an aircrew training scheme had been in the minds of the British Air Ministry for some time. In his 'A Last Call of Empire', John McCarthy writes,

> The origins of the scheme lay...in that cluster of expectations that stemmed from dominion participation in the first world war... Obviously, considerable confidence must have existed in Whitehall that many of the precedents set between 1914 and 1918 would be followed.[2]

Air Vice-Marshal Richard Williams RAAF had been sent on a two year attachment to the RAF in 1939. When he arrived in London he visited the Air Ministry to see the then RAF Air Member for Personnel, Air Vice-Marshal Charles Portal. Williams writes that they 'discussed the prospects of war and he [Portal] stated that if war came the RAF would suggest that both United Kingdom and Dominion personnel be trained in an organization to be set up in Canada and all then serve in the Royal Air Force as the Canadians had done in the 1914-18 war'.[3]

Subsequently, on 26 September 1939, the British Secretary of State for Dominion Affairs, Anthony Eden, sent a cable to the Australian Prime Minister pointing out the 'grave disability' of the Royal Air Force. Eden was asking the dominions to agree in principle to a joint training plan.[4] On 5 October 1939 the Australian War Cabinet agreed to the proposal for aircrew training in the dominions. They sent an air mission to Ottawa to confer with representatives of the United Kingdom, Canada and New Zealand. Australian Prime Minister Menzies told his wireless listeners on 11 October 1939, 'It is no wonder that at this hour of suspense, or of real peril, and of supreme effort, Great Britain should have turned to...the Dominions, and to us perhaps not least of all'.[5]

There seemed to be general support for the plan. An article in *The Bulletin* commented, 'The Governments of Britain, Canada, Australia and Maoriland have become parties to what Mr. Menzies describes as a plan for the building of a vast Empire Air Force which "might well be not only the most spectacular but the most decisive joint effort that will be made by the British nations in time of war". According to the P.M., it will mean that Australia, instead of training and maintaining hundreds of skilled flying personnel, will be training and maintaining thousands.'

Australia now had to choose between its national plan for an expeditionary air force, and joining what would become virtually an Empire Air Force. The original idea for an Australian expeditionary air force was postponed. It was eventually abandoned and the men and resources turned to this huge training task, the Empire Air Training Scheme, which Australia now chose to support. The Scheme linked training organizations in Canada, Australia and New Zealand (and later Rhodesia) and the contracts, known as the Riversdale Agreements (the Scheme was finalized at a Conference chaired by Lord Riversdale), came into force on 17 December 1939. At that stage it was agreed that, once the full organisation was in place, Australia would contribute 10,000 aircrew each year.[6]

From the beginning there was also general agreement that, as far as possible, aircrew of the participating nations should be grouped together. Article XV of the four-party agreement stated, '...that pupils trained in accordance with this agreement shall, after training is completed, be identified with Australia either by the method of organising Australian units or formations or in some other way'.[7] Australian Article XV Squadrons were allocated the numbers 450 to 467, followed by the distinguishing letters RAAF. They were intended to be all–Australian but were to fly under the operational control of the RAF.

Unfortunately, the Australian Government lost the right to post its own nationals serving under the Empire Air Training Scheme, and at times even RAAF Headquarters in London had only a hazy idea of what was happening to RAAF designated squadrons. The composition of Australia's Article XV Squadrons was determined largely by the RAF. By supporting the new Scheme, much of the RAAF became a training organization for the RAF. At the peak of the Scheme in 1943, Australia ran twenty–six out of a total of 333 Air Training Scheme schools and over twenty–seven thousand Australians were trained as aircrew. Australia ended the war with sixteen Article XV squadrons.

There has been controversy over Australia's decision to join the Scheme. At the time it was supported by both political parties. Now, some argue that it seriously prejudiced the possibility of providing for local defence in 1941-42, that it had a disadvantageous effect on the development of the command structure of the RAAF during the period 1942 to 1945, and that it denied Australian air force officers any chance to command large air formations because they lacked experience in higher command. Menzies later commented that the Australian aircrew trained under the Scheme were 'surrendered' to the United Kingdom.[8]

The men who joined the Scheme didn't question strategic or political motives. All they wanted was to do a meaningful job and to be useful operational members of the Air Force. Still, Australia paid a heavy price. Nearly seven thousand Australian airmen were killed while serving with the RAF, over four thousand in Bomber Command alone.

Recruiting was never a problem. The training scheme was entirely voluntary and news of it led many young men to hang back from joining the Army's Second AIF in the hope of entering the air force. Australia's Official History notes that, within a few months of the start of the war, 'the numbers of applicants for enlistment in the air force would soon far out–number those enlisted in the A.I.F.'.[9]

Some men had a struggle joining the Air Force. Bill Waldock from North Perth had been called up into the militia after the War started. He was with the 3rd Field Regiment in Western Australia and he recalls,

> While [the Regiment] was at Narrogin I decided to join the RAAF but the Second–in–Command of the Regiment wouldn't let me go. I paraded myself to the Commanding Officer who gave me permission to go to Perth to do the testing. The RAAF accepted me but the Major wouldn't release me. I started doing correspondence courses with the RAAF anyway. When the Unit was at Dandaragan in Western Australia all the men in our battery were asked to volunteer for the Second AIF. I didn't because I still wanted to join the RAAF. Eventually when our Major was posted out I was released and joined the RAAF.[10]

Bill Waldock had a frustrating time waiting for courses and for a posting to an operational unit. After initial recruit and aircrew training he left Australia in 1943 and finally joined 455 Squadron early in 1945.

There was such a flood of applicants that the RAAF set up a reservist scheme. Men waiting to be called up could start some air force training while continuing their civilian occupations. Peter Ilbery's experiences were typical. He was from Neutral Bay in Sydney and he and his school mate Bill Mitchell put their names down for aircrew training as soon as they were old enough. Ilbery says,

> In 1942 Bill and I were called up. Because there was quite a queue for aircrew training we were recruited as aircrew 'guards' and went to basic training at Parkes, New South Wales. After a few weeks we went to Uranquinty, known as 'Bar Twenty' near Wagga Wagga.[11]

Peter Ilbery had to wait for another four months before he started flying training (and many more months in training before joining 455 Squadron as an operational Beaufighter pilot).

* * *

'This squadron really had two separate foundations', wrote Flying Officer John Lawson.[12] In Australia in May 1941, Lawson reported for duty at Williamtown air base, near Newcastle, New South Wales. He arrived as Adjutant, temporary commander and the sole member of 455 Squadron. The number of personnel posted to the Squadron grew slowly and preparations were made to move to Britain. By the end of June 1941 the Squadron in Australia consisted of two officers and two hundred and thirty-two men, almost a complete establishment of trades for groundcrew to operate two flights of Wellington bombers.

Meanwhile, 455 Squadron RAAF was officially formed in England on 6 June 1941. The first RAAF Squadron to join Bomber Command, it was designated a medium-bomber squadron with Number 5 Group and stationed at Swinderby in Lincolnshire. The Squadron had no crews and had to wait for more than a month before receiving its first Hampden aircraft.

The Australian party left Sydney in July 1941 and would not arrive at Swinderby until September. While the Australians were making their way to Britain an RAF skeleton ground-staff was provided for the Squadron at

Swinderby and the first operation by a 455 Squadron aircraft took place on 29 August 1941. A Hampden bomber, piloted by Squadron Leader Derek French (an Australian serving with the RAF), took part in a raid on Frankfurt. This entry is from 455 Squadron's Operations Record Book:

29 August 1941:
Rain showers 0200hrs, otherwise cloudy with good visibility. Squadron did its first operational flight with Hampden..and following crew. S/Ldr French (Captain), Sgt Pratt (Obs), F/Sgt Maidmont (WOP/AG) Sgt Bernard (A/G. 50 Sqd). Frankfurt was the target. Operation successful and aircraft returned safely. No casualties.

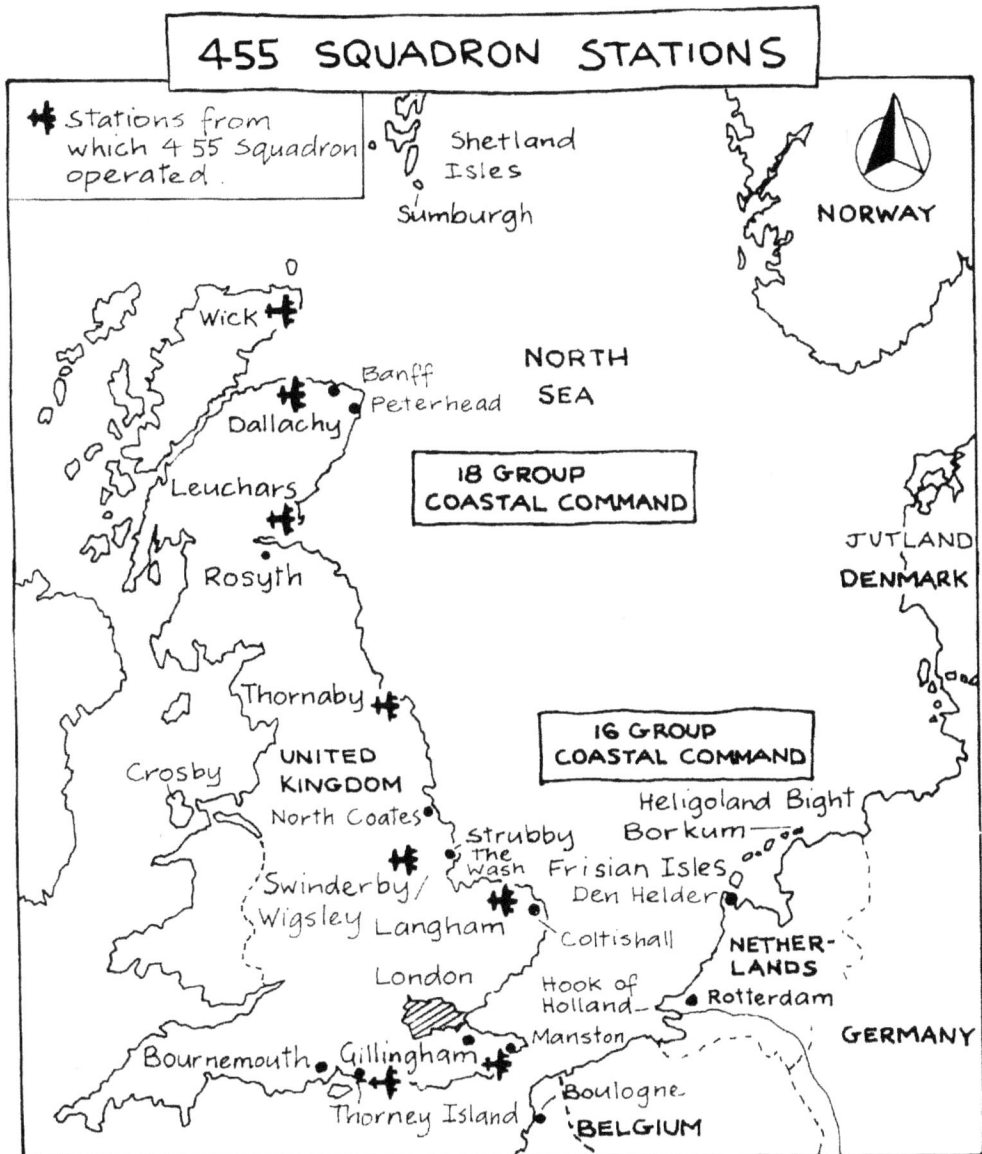

455 SQUADRON STATIONS

⚜ Stations from which 455 Squadron operated.

Shetland Isles
Sumburgh
NORWAY
Wick
Banff
Peterhead
NORTH SEA
Dallachy
Leuchars
18 GROUP COASTAL COMMAND
JUTLAND
DENMARK
Rosyth
Thornaby
16 GROUP COASTAL COMMAND
Crosby
UNITED KINGDOM
North Coates
Heligoland Bight
Borkum
Strubby
The Wash
Frisian Isles
Swinderby
Wigsley
Den Helder
Langham
Coltishall
NETHER-LANDS
London
Hook of Holland
Rotterdam
GERMANY
Bournemouth
Gillingham
Manston
Thorney Island
Boulogne
BELGIUM

5

'You had to...be prepared to throw it around': Handley Page Hampden UB-A, AD 739 of 455 Squadron. 455 Squadron operated with the Hampden in RAF Bomber Command from August 1941. The Hampden bomber carried a crew of four (pilot, navigator, wireless operator-air gunner and air gunner) and was powered by twin Pegasus engines. (Photograph courtesy C. Bastion)

The new Squadron faced many difficulties. There were no operationally experienced pilots in the Squadron except for the new Commander, Wing Commander Gyll-Murray DSO, RAF, and his two squadron leaders, French and Reynolds. There was an acute lack of aircraft and aircrews, and a Squadron that was supposed to be Australian was entirely without RAAF aircrew. During October more aircraft began to arrive and by the end of the month the Squadron had over twenty Hampden bombers. 455 was the first RAAF Squadron to become operational under the Empire Air Training Scheme.

The Handley Page Hampden was a medium bomber. Its twin Pegasus engines gave it a cruising speed of 130 knots and, for its size, it could carry a heavy bomb load. Its armament was six gas-operated .303-inch machine guns. The crew of four consisted of a pilot, navigator, wireless operator-air gunner and air gunner. Its appearance gave it the nickname 'The Flying Tadpole' or 'The Flying Suitcase'.

455 Squadron began regular operations from October 1941. On 7 November 1941, seven Squadron aircraft took part in a diversionary attack on Cologne while a larger force of Wellingtons, Stirlings and Manchesters attacked Berlin. Two of 455's Hampdens did not return - those captained by Pilot Officer Tony Gordon and Sergeant M.A. Jenkins RAF. Gordon's crew became prisoners of war, but Jenkins and his crew were killed.

John Lawson noted that 'The year 1941 closed very cheerfully. The Squadron was practically up to full strength. No casualties had been suffered in December in spite of quite extensive operations and the year had closed with two highly successful raids. The Squadron was still a League of Nations affair but none the worse for that.'[13] Recent postings had brought the Australian air crew numbers almost, but not quite, up to RAF numbers. There were also New Zealanders, Canadians, Rhodesians, and a representative of Kenya.

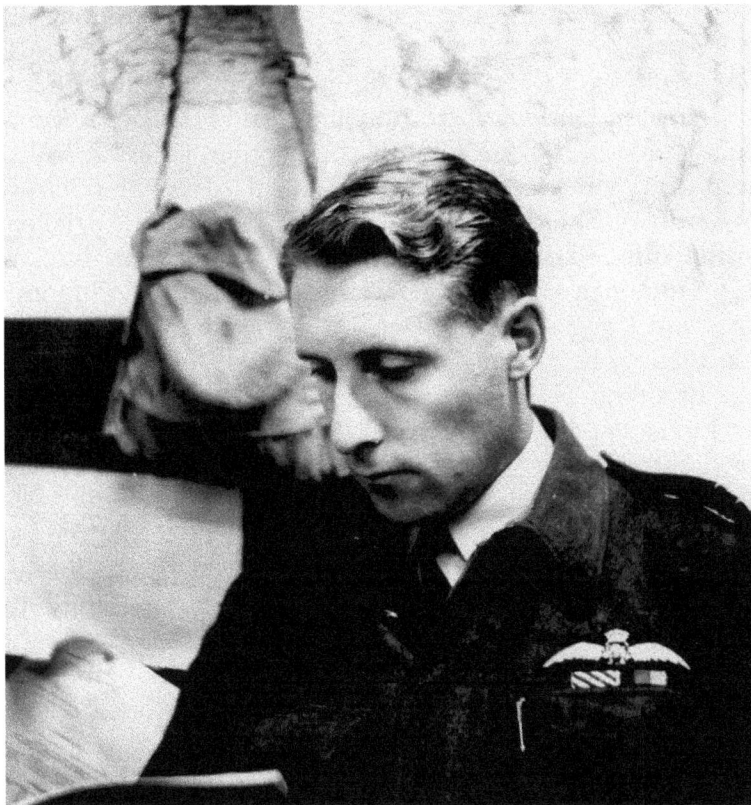

'At last the Squadron was to have an Australian Commanding Officer': Wing
Commander Grant Lindeman DFC from Sydney. Grant Lindeman trained at
Point Cook with the RAAF and was offered a short service commission with the
RAF before the War. He remained with the RAF when the War began and
commanded 455 Squadron from January 1942 until February 1943. (Courtesy
G. Lindeman)

In 1941 the Battle of Britain had only just been won and new aircrew were
needed urgently in many RAF units. Australian aircrew arriving in Britain
were sent first to Number 11 Personnel Dispatch and Reception Centre at
Bournemouth, a south coast resort centre, where many large hotels had been
requisitioned. The Reception Centre was later moved to Brighton in May
1943, and many Australian aircrew remember the Hotels Metropole and
Grand, then stripped of luxuries. Aircrew were then posted to operational
training units and finally to operational squadrons. Because the posting section
at the Reception Centre at Bournemouth was concerned mainly with filling
vacancies at RAF operational training units, six months after its establishment
455 Squadron held only a token number of RAAF crews. The slow progress
in establishing Australian squadrons overseas was criticized in Australia by the
press and by the Federal Opposition in Parliament. Australia's Official History
of the Air War notes that Mr J.A. Bessel, Minister for Supply and Shipping,
publicly deplored the fact that Australian airmen were scattered over many
squadrons instead of being concentrated into national units. The Minister for
Air, Mr J. McEwen, replied that the Government's ultimate aim was to

achieve fully-Australian squadrons but that, 'a large measure of dilution was inevitable at first.'[15]

And at last the Squadron was to have an Australian Commanding Officer. Wing Commander Grant Lindeman was from Sydney, and he had trained as an air cadet at Point Cook in 1934-35. With three other pilots from his course he was offered a short service commission with the RAF. Lindeman was posted to 500 Auxiliary Squadron at Manston in Kent, flying Vickers Virginia bombers which differed only slightly from the Vickers Vimy used in the First World War. Lindeman says that, 'Even in those days the "Ginnies" were museum pieces and I got a great deal of fun flying them'.[14] Then, he flew troop carriers at Heliopolis, in late 1935, and over the next three years piloted a wide range of aircraft. His short service commission ended just as the War started, and Lindeman elected to remain with the RAF (although he wore the 'Australia' flash on his RAF uniform). He took command of 455 Squadron on 3 January 1942.

455 Squadron's time in Bomber Command lasted less than twelve months. A dramatic event in the English Channel early in 1942 showed up weaknesses in the RAF which eventually led to 455 Squadron being transferred to Coastal Command.

In February 1942, the German naval squadron sheltering in the harbour at Brest, in France, was threatening to break out in order to return to Germany. The powerful German battle cruisers *Scharnhorst* and *Gneisenau*, and the cruiser *Prinz Eugen* had been undergoing repairs in Brest and it appeared that they would be likely to attempt to force their way back to Germany up the English Channel. A plan to intercept them involving Bomber, Fighter and Coastal Commands was prepared. For the Germans, the Channel route was their only real hope of returning these ships to Germany. Hitler had said, 'You can count on this; from my previous experience I do not believe the British capable of the conception and execution of lightning decisions such as will be required for the transfer of their air and sea forces to meet the boldness of our operation'.[16]

On the night of 10 February 1942 the German warships sailed. They eluded a British submarine on patrol outside the harbour and jammed English coast-watching radar stations. They ran almost the whole course of the English Channel before they were discovered, on 12 February, by an air patrol off Boulogne. There was considerable confusion because widely differing position reports were received by Bomber Command and Coastal Command.

Coastal Command attacked the German naval squadron but made little impression. Bomber Command joined in. From their station at Wigsley (a satellite airfield for Swinderby) nine 455 Squadron Hampdens, including those aircraft flown by 455's Commanding Officer Grant Lindeman and Flight Lieutenant Walter Perrin, set off at short intervals to the target area. Perrin, an Australian serving with the RAF, had trained at Point Cook, Victoria, with the Citizen Air Force, and in 1926 he qualified as a pilot. When the War broke out Perrin applied for flying duties but was told he was too old and he was offered an administrative job. He left Australia and worked for a time as a metallurgist in South Africa. In May 1940, he took a job in the stokehold of a ship to get from Capetown to Britain, to join the RAF. Without his log book and papers he had to go through all his flying training again. The RAF rules for the maximum age of operational pilots were relaxed and Perrin was given operational training. He joined 455 Squadron in December 1941.

On the morning of 12 February, conditions made success practically impossible. The cloud base was at about one thousand feet, the target was moving on an unknown course and anti-aircraft fire was exceptionally heavy. Of the 239 aircraft of Bomber Command that took off on this operation, only thirty-nine, including five Hampdens of 455 Squadron, located and attacked the targets. After Perrin had taken off from Wigsley there were no signals from him and he failed to return, presumed lost in the target area. It was Perrin's eighth operational sortie.[17]

The British air and naval attack had failed to stop the German warships. This was a bitter disappointment. It was another considerable set-back to British and Royal Navy prestige, already battered by the successes of the U-boat in the Atlantic, by the sinking of the *Prince Of Wales* and the *Repulse*, and by the Japanese occupation of Malaya. There was an immediate Inquiry into the dash up the Channel by the German squadron. Wing Commander Grant Lindeman attended the Inquiry as a witness, along with one of his flight commanders, Squadron Leader Jimmy Clift. Lindeman recalls,

> The President [of the Board] was a silvery-haired old Admiral in a black alpaca coat. I had already explained that we were armed with armour-piercing bombs which had to be dropped from 7,000 ft or more to get up sufficient speed to pierce armour. As the cloud in the target area was 10/10ths at 1,000 feet our bombs (if they hit!) would have merely bounced off the ship's armour. The dear old gentleman asked Clift, 'When you saw that your bombs were ineffective, why did you not fly your aircraft into the bridge of the ship?' With a perfectly serious face Jimmy replied, 'The idea never occurred to me, Sir,' which caused the attending Air Vice-Marshal to snigger.[18]

The Inquiry highlighted the limitations of the British air and naval forces. At that time, Coastal Command's main torpedo force was employed in the Middle East; its flying strength in Britain had been reduced to two and a half squadrons.[19] The escape of the *Scharnhorst* and *Gneisenau* also revealed short-comings in control, co-ordination and training of aircrew engaged on anti-shipping strikes. The incoming Air Officer Commanding Coastal Command, Air Marshal Joubert, proposed that his Number 16 Group should assume direction of the anti-shipping campaign and that Bomber Command should be made responsible for all mine-laying operations in enemy waters, thereby 'releasing Coastal torpedo squadrons from a task that had hindered training for their primary role'.[20] The Inquiry also reported that 'The need for the development of a powerful and highly-trained striking force of torpedo bombers seems to be one which calls for urgent consideration'.[21] However, this general clearing of the air did not bring about immediate success. Coastal Command was still beset by shortages of aircraft and aircrew, and Joubert believed that with the limitations of its existing resources, Coastal Command could not operate efficiently. As a result, 455 Squadron and 144 Squadron RAF were to be transferred from Bomber Command to Coastal Command, as torpedo-bombing squadrons.

Before that transfer, 455 Squadron experienced a last important and very intensive run of operations with Bomber Command. One notable sortie took place on the night of 1 April 1942, when Wing Commander Grant Lindeman led a low-level train hunt over north-western Germany. He took with him the Squadron's Gunnery Officer, Flight Lieutenant Findlay Flemming RAF, a veteran of over one hundred operational hours. They took off from Wigsley at just after 8 pm, followed immediately by a second Hampden flown by Flying Officer Roy RAF. There was a full moon that night - pilots said it was almost like daylight, and they found that they could fly safely as low as two hundred feet. Lindeman recalls the flight to the target,

> On the flight across Holland it was bright moonlight and the ground was white with snow. We flew very low to avoid fighters and it was encouraging to see the 'V' sign, made by flashlight from a lonely farmhouse. At one point we were picked up by searchlights in spite of our low altitude, but I evaded them by flying even lower so that we were masked by a line of trees.[22]

Nearly two hours out from Wigsley, Lindeman found the Rheine to Munster railway line. There were four trains on the line and Lindeman led in to attack. He says,

> We were only at a few hundred feet, bombing and machine-gunning troop trains, when from carelessness or excitement I did a steep turn with too little bank and the aircraft skidded sideways with the rudders locked hard over. I had to twist the aircraft out of it by brute force, full throttle on the inner engine and throttle right back on the outer.[23]

He then flew on to the Rheine-Burghsteinfurt line and from nine hundred feet dropped his bombs onto the tracks. Lindeman's gunner Flemming found the operation exciting and troublesome - troublesome because of the air sickness brought on by the violence of Lindeman's evasive action. Lindeman continues,

> On the way home, as we approached the coast I climbed for height for crossing the sea but just as we crossed the coastline two flak batteries, one on each side of us, caught us in a cross-fire on tracer. We dived for sea again as fast as we could.

Lindeman had his Hampden and crew back at Wigsley at 2.13 am next morning, after an operation lasting over six hours.

The move to Coastal Command was not popular with 455 Squadron. The work of Coastal was not understood, nor even highly regarded, by some in Bomber Command. Friendships were broken up and they left behind in Bomber Command some great Squadron personalities, including 'Mickey' Martin, Lindeman's navigator, Bob Hay (both selected by Guy Gibson for the Dambusters raid), and Jimmy Clift.[24] Farewell parties were held at Swinderby on the nights of 24 and 25 April 1942, and the next day 455 Squadron made its way to its new home at Leuchars, Fifeshire, in Scotland.

Coastal Command had been born out of a major pre-war reorganisation of the RAF. In 1936 the RAF had begun to expand and the whole service was re-organized into four commands: Fighter, Bomber, Coastal, and Training. Each command headquarters exercised its control through a number of group headquarters, which in turn ran the squadrons and detached flights. This was the basis of the command system used in Britain throughout the Second World War.

At the start of the War, the Commander-in-Chief of Coastal Command was Air Marshal Sir Frederick Bowhill. He moved his headquarters from the south coast of England to a private house at Northwood, Middlesex, close to the equivalent headquarters of the other services. Bowhill set up a combined headquarters where the navy, army and air force staff worked together. In July 1939 all units in Coastal Command had been given their mobilisation orders. Coastal Command had a triple task: 'find the enemy, strike the enemy, protect our ships'. It tasked its Groups,

15 Group - with headquarters at Plymouth on the south-west coast of England, to conduct anti-submarine patrols on the South West approaches to Britain.

16 Group - with headquarters at Gillingham, near the mouth of the Thames, to conduct anti-shipping patrols in the English Channel and along the Dutch Coast.

17 Group - with headquarters at Gosport on the south coast of England, to carry out operational training on Coastal Command aircraft.

18 Group - with headquarters at Rosyth in Scotland, to carry out anti-submarine and anti-shipping patrols in the North Sea.

Swedish iron ore was of particular interest to Coastal Command. At the outbreak of the War, Germany's total overseas imports were twenty-nine million tons. Most important for their war effort was the eleven million tons of high grade iron ore imported each year from northern Sweden. Germany's iron and steel industry was the largest in the world after the United States, and its industry was geared to the use of this Swedish ore. It was the need to ensure a continued supply of raw materials, especially the high grade iron ore, that prompted Hitler's invasion of Denmark, Norway and the Low Countries in the spring of 1940. Grand Admiral Raeder, the Supreme Commander of the German Navy until 1943, considered that without that Swedish ore, Germany would not have been able to keep its armaments industry going.[25]

Naturally, Norway and her coastline were vital to Germany. The iron ore was mined in northern Sweden and one of the most important transport routes to Germany was via the ice-free port of Narvik in Norway. The ore was then shipped down the Norwegian coast to Rotterdam in Holland and sent by barge up the Rhine to the steel factories in northern Germany. Along much of the sea route down the Norwegian coast, numerous off-shore islands afforded shipping some protection from attacks by Allied submarines or light surface forces, but not from air attacks.

At the end of June 1940, Germany was in possession of the whole coastline of western Europe from North Cape at the northern tip of Norway to Bordeaux in France. Coastal Command had less than five hundred aircraft and was overloaded by new tasks. It was heavily involved in anti-invasion operations; it lent valuable support to Fighter Command in the Battle of Britain and while continuing its North Sea patrols, it had to attempt to keep a close watch on the Norwegian coastline with its maze of fiords and channels. It was called on for coastal reconnaissance; to provide long range protection for shipping, and for anti-U-boat measures. The Command's most pressing needs in 1941 were 'numbers of long-range aircraft and lethal weapons'.[26] Fortunately, new aircrew and new aircraft were now being made available.

On joining Coastal Command, 455 Squadron was changed significantly, with a new base, new weapons, new crews, a new Command and new tasks. The Squadron was to operate only two Hampden flights instead of the three it had in Bomber Command. The Commanding Officer of 455 Squadron, Wing Commander Grant Lindeman, had been allowed to take with him to Leuchars only two experienced crews and ten junior crews to form a torpedo strike squadron while many of the Squadron's Australian groundcrews and air crews remained in Bomber Command.

455 Squadron joined a busy base at Leuchars. The Station was near a small village on the east coast of Scotland within easy reach of Dundee and St Andrews, with its famous golf course. This RAF station was well established and had been a Royal Navy air base during the First World War. At the Station was a Photographic Reconnaissance Unit, a Gunnery Training School, 42 Squadron RAF flying torpedo carrying Beauforts, and both 144 Squadron RAF and 455 Squadron under training in their Hampdens.

Jack Davenport was an RAAF Flight Lieutenant and a deputy flight commander in 455 Squadron during its early days at Leuchars. He was from Blakehurst, New South Wales, had been at a Sydney high school before the War, and joined the militia with the 30th Battalion. He transferred to the Air Force at the end of 1940 and, through an unusual run of luck and some very good performances, was on his first operational training sortie in just over ten months after joining the Air Force. Davenport was posted to 455 Squadron in March 1942 as a Pilot Officer. As captain of a Hampden aircraft he took part in nine sorties within twenty days, some against the most heavily defended targets in Europe, including Hamburg, Essen and Dortmund. Davenport became an important Squadron personality and was one of its most famous commanders. He enjoyed his Hampden days and recalls,

> Early in our time in Coastal Command we flew the Hampden bomber. It was a lovely aircraft to fly. You had to not be afraid of it and be prepared to throw it around. When we were stationed at Leuchars in Scotland we used to fly over an RAF Spitfire squadron in Edinburgh and they would scramble and try to 'shoot us down' with their cameras. We usually survived because of our manoeuvrability and slow speed. Unfortunately, the Hampden had a reputation for its tendency for developing a stabilized yaw (when the twin rudders locked on if the aircraft skidded in a turn). At one time during my time at the Operational Training Unit I got into a spin and the crew had to bail out. One of the crew was trying to open

the rear door and pulled on the door strut instead of the release bar. He managed to bend the strut which was about half inch diameter steel. When we got back we showed him what he'd done. On the ground and out of danger he couldn't budge it.[27]

At Leuchars, 455 Squadron began modifying its Hampdens and retraining its crews for operations with torpedoes. Jack Davenport was closely involved in the retraining program. He says, 'When we joined Coastal Command we were faced with a number of problems. One was how to successfully drop WWl torpedoes from planes going much faster than the old Swordfish.'[28]

The torpedo supplied for the Hampden had been developed by the Admiralty during the First World War. Launched by the Hampden at between 120 and 130 knots, it would slow down to forty knots in the water and evasive action by enemy shipping was usually very effective. The Hampden pilots had to learn to fly in tight formations, at very low altitudes, and to drop torpedoes very accurately. Grant Lindeman recalls that 455 Squadron had hoped to learn about torpedo tactics from an experienced Beaufort squadron stationed nearby, but 455 were 'dismayed to learn that they had no considered plan of action other than to arrive together at the target, after which it was every man for himself, to circle and drop his torpedoes as the situation developed.'[29]

Finally, in July 1942, 455 Squadron was declared fit for torpedo operations with Coastal Command. Jack Davenport recalled the early operations from Leuchars,

> After a lot of training we did a number of operations. The problems we'd had with the torpedoes were highlighted when we were sent to attack a light cruiser sighted off the Norwegian Coast. We were about 70% trained by then, and it was before our intelligence system had been fully developed to the remarkable extent later achieved. The torpedoes had to be dropped at about 800 to 1,100 yards, and 'aimed off' to allow for the speed of the boat. After we released the torpedo, all the enemy ships had to do was to change course towards the attacking aircraft, providing a small target, and the torpedo would likely miss. That required the design of special sights, which we made from bits of wood and nails. On top of that, the torpedo had to hit the ship under the water line, the deeper the better. Since each class of enemy ship had different under-water dimensions, we had to know what class of ships we were attacking before we left for the strike as the depth setting for the torpedoes could not be altered after take-off. This meant that our intelligence was very important for success in strikes.[30]

455 Squadron spent July waiting for a chance to get involved in more operations. That chance came soon - in Russia.

There are few more remarkable periods in the Squadron's early history than its expedition to Russia in September 1942. At that stage, success of anti-shipping strike activities of Coastal Command had fallen away, in some ways because of the unsuitability of the Hampden aircraft and its weapons. Coastal Command now put into operation a plan to give special protection to merchant convoys travelling from Scotland to Murmansk, Russia. Two squadrons of Hampden torpedo bombers were to be stationed in Russia to deter Norway-based German warships from leaving harbour, and to attack these warships at sea should they sail against the convoys. 455 Squadron teamed with 144 Squadron RAF for the task.

The planning was supposed to have been conducted in secrecy. Groundcrews sailed for Murmansk in July 1942 and Jack Davenport recalls that, early in the preparation phase, 455 Squadron personnel were issued tropical clothing so no-one would know the destination. 'Of course the moment we were issued with that kit we all knew where we were off to.'[31] The convoy that they were to protect, designated PQ 18, sailed for Russia on 2 September 1942. That same day the squadrons left their base at Leuchars for Sumburgh airfield in the Shetland Islands.

The departure from Leuchars was an emotional time. Few of the crews leaving can really have expected a safe return. 144 Squadron took off first and as the flight to Sumburgh was relatively short, they put on an alarming display of aerobatics over the airfield. 455 Squadron's take-off was regulation, preceded by Wing Commander Lindeman's instructions over the radio, '455 will take off like gentlemen'.[32]

The two squadrons left Sumburgh on 4 September. Jack Davenport remembered,

> The runway at Sumburgh was quite short with a cliff at one end and the sea at the other. We pulled each aircraft up against the cliff at the end of the runway, topped them up with fuel and set off. Our planned trip was beyond the normal endurance of the Hampden.[33]

The aircraft took off at the rate of about one per minute. One of the 455 Squadron flight commanders, Squadron Leader Jimmy Catanach from Melbourne could not restrain himself. Wing Commander Lindeman recorded in his diary,

> he taxi-ed into the first gap in the line and was off like a blooming rocket. I've never seen such a wealth of superfluous energy in any individual over the age of twelve as Jimmy constantly had at his disposal. He didn't drink or smoke; he talked at an incredible speed; he couldn't stand still for a second, but hopped all the time you were talking to him till you were nearly giddy. He was a most excellent Flight Commander, and was probably the most generally liked man in the whole Squadron.[34]

The plan was to fly from Sumburgh over the North Sea, to Norway, Sweden and Finland then to North Russia. First landfall was to be Afrikandar near the Gulf of Bothnia, south of Murmansk. From Afrikandar there would be a relatively short flight to Vaenga, an airfield near Murmansk.

The planes flew through the night and crossed the Finnish border at dawn. They arrived in the vicinity of the landing field to find the whole area covered in mist. Wing Commander Lindeman had to find a way to land his Squadron,

> It was only about two hundred feet thick, but it looked quite solid...It was going

to be a bit tricky getting down through that stuff...We noticed several other Hampdens hunting around for a place to perch, so we decided it was worth a try before going away up north again. It wasn't a pleasant feeling to be looking for an unknown aerodrome under a sheet of fog, in a strange, wild country so very far from home, and with not much petrol to play about with.

Eventually, Lindeman saw a green Very light shoot up from the ground. He caught sight of a flat sandy landing ground through a break in the fog, but his landing was as hazardous as the long flight. Lindeman said,

> You could see the ground through the mist if you were high enough up, but as soon as you got down low you lost sight of the ground, and of course once you were in the fog you couldn't see a thing at all. The only thing to do was to approach fairly high and once in the mist to keep the aircraft in a straight glide by instruments until the ground could be seen beneath the wheels. I held the aircraft at that height with the engines until the airfield boundary flashed by, then throttled back and let it settle on the ground.[35]

It was an extraordinary flight of endurance and navigation. Fourteen of 455 Squadron's sixteen aircraft landed safely in Russia. Jimmy Catanach force-landed in northern Norway and he and his crew were taken prisoner by the Germans. Another 455 Squadron aircraft crashed off the coast of Sweden. Sergeant Edward Smart and his crew were killed.[36] Jack Davenport was forced to land by two Russian Hurricane fighters on a mud strip called Monchegersk. Davenport says, 'The Russians had never heard of a Hampden aircraft nor of our visit.'[37] 144 Squadron RAF did not do so well. They lost six of their sixteen crews on the way to Russia.

As convoy PQ 18 sailed, the Germans were preparing to do their utmost to repeat the damage they had inflicted on earlier convoys. On 10 September 1942 the German warships *Admiral Scheer, Hipper* and *Koln* moved north from Narvik to Alten Fiord in the very north of Norway. However, Hitler had insisted that risks should not be taken with his heavy ships and Admiral Raeder cancelled their intended attacks. They stayed in harbour throughout PQ 18's passage. The Luftwaffe, personally urged on by Goering, attacked the convoy in strength. The air battle lasted with scarcely a break until PQ 18 reached the entrance to the White Sea. In all, convoy PQ 18 lost thirteen merchantmen – ten to air attacks and three to U-boat torpedoes. Twenty-seven merchantmen arrived at their destination.

Jack Davenport recalls the operations from 455 Squadron's base in Russia,

> We only did one operation from Vaenga. We got news that the German warships were moving from Alten Fiord. I led the operation because Grant Lindeman, our very capable Commanding Officer had been injured when he threw away a piece of metal that turned out to be a detonator. It was a most frightening and hair raising operation. We had to fly at low level and the strong turbulence made it dangerous. We sighted nothing and by the time we returned most of the crews were ill. We think the German ships must have had some notice of our approach and returned to their anchorage. During the stay in Russia the aerodrome and our barracks were regularly bombed by the Germans flying from a short distance away

455 SQUADRON RUSSIAN EXPEDITION 1942

BARENTS SEA

Convoy → Route →

75°N

Hampdens

Hipper
Köln
Admiral Scheer
4 destroyers

Tirpitz
6 destroyers

North Cape

Alten Fiord

Murmansk Vaenga

Narvik

FINLAND

SWEDEN

NORWAY

WHITE
SEA

65°

U.S.S.R.

63°N

455 Squadron

12°E   20°   30°   40°   48°E

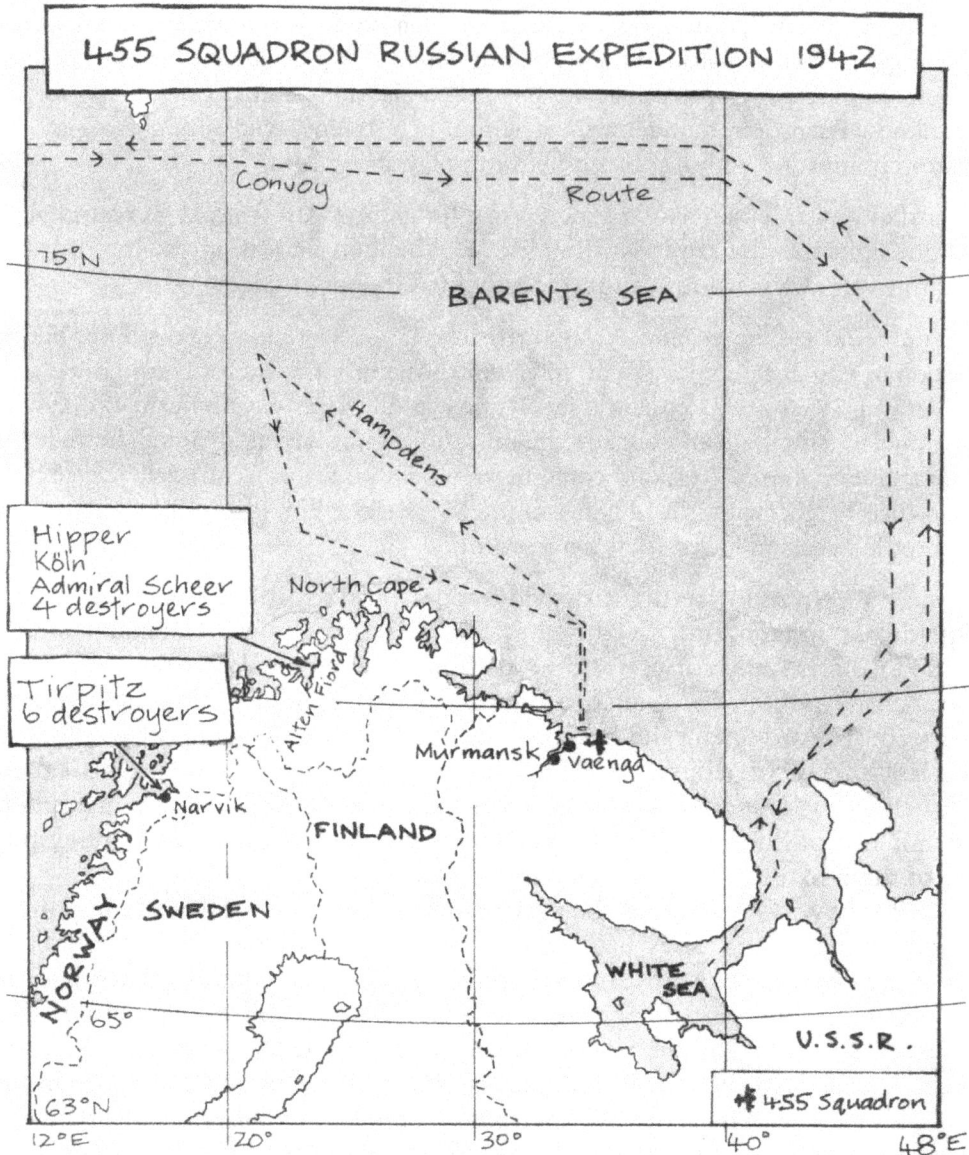

in northern Finland. The return flight, against prevailing winds, was well beyond the range of the Hampdens. The aircraft were given to the Russians who were taught to fly them mainly by Flying Officer Dick Humphrey. Their pilots were good but rough. We found it hard to watch their landings.[38]

The operation to protect the convoy from surface attack had been a success and 455 Squadron received a message of congratulations from Winston Churchill. The men returned to Scotland by sea in late October 1942. After the return some of the crews were interviewed by the BBC. The BBC interviewer said to Grant Lindeman, 'I suppose you'd be busting to get back up there?' Lindeman replied, 'Should we be given the opportunity to return, we will unselfishly stand aside and give others a turn'.[39]

New Hampdens and crews arrived at Leuchars in November to replace those left in Russia. Jack Davenport was promoted to Squadron Leader and given command of 'A' Flight, in place of the captured Jim Catanach.

Jim Catanach was held prisoner at Stalag Luft III, where the standard of escaping knowledge was very high. Nearly one hundred tunnels were begun by prisoners between March 1941 and March 1944. In 1943 the prisoners' escape committee began work on a grand design by which no fewer than 200 airmen were to escape at one time. Three tunnels were begun simultaneously and named Tom, Dick, and Harry. Harry was to be the real escape route, and although the actual escape had been originally planned for summer, increasing probes by the German internal security police led to a decision to break out when the moon conditions were favourable on 24–25 March 1944. Twenty Australians, including Catanach, joined other escapees in the hut containing the entrance to the tunnel. Seventy-six men had escaped by 5 am next morning when a guard saw the tunnel mouth. Hitler ordered that more than half of the men recaptured were to be shot as a deterrent to all other prisoners. Fifty of the recaptured men were killed. One of these was Squadron Leader Jim Catanach.

In December 1942, 455 Squadron's daylight patrols along the Norwegian Coast began again. Weather conditions were almost always bad and success was limited. The task was made more difficult because of the Hampden's vulnerability to German shore-based defences and to fighter aircraft. This problem existed everywhere in the Command; in the first four months of 1942 only six ships were sunk by Coastal at a cost of fifty-five aircraft. By June that year it had been estimated that, over the preceding three months, one in four of all attacking aircraft had been lost. Coastal Command standing orders forbade Hampden squadrons to attack unless there was adequate cloud cover. In addition, the German merchant ships had improved their defensive armament and the number of escort vessels was increased. These problems were eventually solved with the evolution of strike wings.

Strike wing tactics had originated in Malta in mid-1942. There, some experiments had been carried out combining anti-flak aircraft and torpedo bombers with good results. The idea was for a squadron of fast torpedo bombers to be accompanied by one or more squadrons of aircraft with equal performance, that used cannon fire and bombs to subdue enemy anti-aircraft fire. These escorts would also engage enemy fighters at the critical moments when the torpedo bombers made their deliberate straight and level approach to their targets. The strike wing tactics were perfected in Coastal Command with the Bristol Beaufighter and Air Marshal Joubert, commanding Coastal Command, recognized the potential of this aircraft. He wrote,

> By the end of 1941 the Beaufighter was coming into operation in Coastal Command. My ambition was to form a strike squadron of Torbeaus (torpedo-carrying Beaufighters) supported by at least two Beaufighter Squadrons armed, if possible, with rockets as well as cannons. With a wing this sort I felt that real damage could be done to German coastal shipping.[40]

In June 1942 Joubert persuaded the Air Ministry to authorize the modification of Beaufighters as torpedo carriers.

'The result of a remarkably rapid evolution': A Bristol Beaufighter Mk TFX (Torpedo Fighter Mk X) with Invasion Stripes used to identify Allied aircraft during the D-Day invasion period. This Beaufighter is on the apron at RAF Station Langham and carries a pair of 500 pound bombs for use against German E-boats. The nose camera and the navigator's hatch on the underside of the aircraft can be seen.

The Beaufighter was the result of a remarkably rapid evolution of the Bristol Beaufort bomber. The Beaufort had been ordered into production for the RAF in 1936 as a torpedo-bomber and general purpose reconnaissance aircraft for Coastal Command. By 1938 it was apparent to the Bristol company that the RAF was likely to be called upon to go to war with no adequately armed fighter. The head of Bristol's design team, Leslie Frise, suggested that a fighter derivative of the Beaufort bomber would fill the gap, with the Hercules engine to replace the Taurus used in the Beaufort.

The Hercules engine was a fourteen cylinder, two-row radial, air-cooled, sleeve valve engine, rated at 1,410 horsepower. That gave the Beaufighter a top speed of 492 kmh (306 mph) (compared to the production version of the Spitfire with the Rolls Royce Merlin engine, which had 1,030 horsepower and a top speed of 570 kmh [355 mph]). The larger propellor size of the Hercules engines on the Beaufighter required the shortening of the aircraft's nose so that it ended behind the plane of the propellors. That change was one of the most obvious external features of the Beaufighter design.

Bristol suggested a two man crew. The second man was to be located well behind the pilot in the fuselage where he could load the four 20 millimeter cannons that were mounted low in the nose.' The cannon ammunition was held in four boxes, each with space for 240 shells which fired a mixture of solid armour piercing heads, tracer and explosive head. The second crewman was also to act as radio operator, navigator and rear gunner.

Bristol was looking to speed production. They decided to make maximum use of the existing Beaufort jigs, so the full fuselage length of the Beaufort was retained. The rear end of the fuselage, the tail, the outer wings and tail wheel were all to be standard Beaufort parts. It was only a few days after the Beaufort bomber made its first flight that a preliminary layout for a 'Beaufort Fighter' was submitted to the Air Ministry. It brought an enthusiastic response. Bristol was asked to produce four prototypes and soon after, the name 'Beaufighter' was accepted because of its Beaufort origins. The first Beaufighter was flown on 17 July 1939, forty-eight days before War was declared. By the time the initial production order of three hundred aircraft was confirmed, the first prototypes were ready to fly. This was only six months after the first layout was drawn.

Flight testing of the Beaufighter had showed little need for modifications. To provide for possible weight growth and to absorb heavy landing loads, a new Lockheed undercarriage was used. The first operational Beaufighters were delivered to RAF night-fighter squadrons in September 1939. At the end of the Battle of Britain, in October 1940, Beaufighters Mk IF (the F denoting Fighter Command) were being delivered. By the end of the German *Blitz* in May 1941, five RAF home-based night-fighter squadrons (numbers 25, 29, 219, 600 and 604) had been re-equipped with Beaufighters. Early in 1942, plans were made for the Beaufighter to be modified to carry a torpedo, regarded as a vital weapon in the battle against the enemy's sea forces. An aircraft was modified to carry a torpedo and it was also fitted with air brakes to help the pilot maintain a steady run-up to the target. The Torbeau (torpedo Beaufighter) prototype went to the Torpedo Development Unit at Gosport in May 1942.

The production-standard Beaufighter with Coastal Command was designated the Mk IC (C denoting Coastal Command). At the end of 1942, the full production standard became the TF (Torpedo Fighter) Mk X and by the end of the war production of the Beaufighter TF Mk X totalled 2,205. The interval from formulation of the requirement to start of deliveries had been nineteen months, one of the shortest periods recorded for any of the principal aircraft types developed in Britain and used in the Second World War.

The first Coastal Command Beaufighter strike wing, based on the Malta model, was formed by 16 Group in November 1942. Three Beaufighter squadrons, 143 Squadron RAF, 236 Squadron RAF and 254 Squadron RAF combined at North Coates near the coast of Lincolnshire. One was armed with cannon, machine guns and bombs, and the other two with cannon and torpedoes. The new strike wing worked on the principle that up to twelve aircraft should carry torpedoes, depending on the composition of the convoy, and that three anti-flak aircraft were necessary to deal with each flak ship.[41] The two attack waves were to be not more than a minute apart with the ideal interval to be 2000 metres between the anti-flak and the torpedo aircraft. The training for combined wing strikes was to concentrate on quick take-off and forming-up, on rendezvous of wings, and on navigating in large numbers over the sea.[42]

In February 1943, Air Marshal Sir John Slessor succeeded Joubert as Air Officer Commanding Coastal Command. In Coastal Command's monthly confidential journal, the Coastal Command Review, Slessor described the strike wing tactics this way,

> Each strike was a carefully planned, highly organized and hard-fought action involving up to thirty or more aircraft, each of which had its own particular role in the attack. The torpedo aircraft went straight for the merchant ships in the convoy while their escorts dealt with the enemy's escorting ships with bombs, cannon-fire and rockets.[43]

Early in 1943, the future direction of the War was determined by the Allies. At a conference at Casablanca in Morocco, President Roosevelt and Prime Minister Churchill reaffirmed that the war against Germany was to have priority in all Allied efforts. The combined Chiefs of Staff Directive stated that the main Allied aim was 'the progressive destruction and dislocation of the German military, industrial and economic system, and the undermining of the morale of the German people to a point where their capacity to resist is fatally weakened.'

At the beginning of 1943, 455 Squadron's contribution to this strategy was being frustrated by severe weather. In February, Wing Commander Bob Holmes, from Perth, Western Australia, succeeded Lindeman. Lindeman, an enormously respected leader, went on to take new responsibilities in the RAF.[44] In March, plans were made for a return to Russia, but the trip was called off. During strikes in April, 455 Squadron Hampdens began to be escorted by Beaufighters from 235 Squadron RAF. The Beaufighters flew top cover for the Hampdens, diverting the attention of the anti-aircraft fire. This technique was first tried from Leuchars on 12 April 1943 and the presence of the Beaufighters certainly saved some of the Hampdens when they were attacked by four Focke-Wulf Fw190s and two Messerschmitt Bf109s off the Norwegian Coast.

Patrols and strikes by the Squadron continued during May and June 1943, and in August a number of ships were sunk. At the end of August thirty-eight crew members were grounded on completion of their operational tours. This included nearly all of the aircrews that had transferred with the Squadron from Bomber Command. Twenty-three were Australians and they were posted back home. The Commanding Officer of 455 Squadron, Wing Commander Holmes, and the 'A' Flight Commander, Squadron Leader

Clarke DFC stayed in their positions of command. Squadron Leader Jack Davenport had completed his first operational tour and was posted to Turnberry in Ayrshire as Chief Flying Instructor of Number 1 Torpedo Training Centre. There were no operational flights in October because of the poor weather and in November and December 1943, patrols were uneventful.

1943 had seen almost unbroken success for the Allies in Europe. In the Atlantic the U-boat had been temporarily defeated and the Mediterranean had been reopened. In the air, the Western Powers were dominant and the strategic bombardment of Germany had reached a new peak with the introduction of long-range American fighter aircraft. That made it possible for the Allies to attack any target in Germany with a thousand bombers by day as well as by night. In Australia, the face of the War had changed a great deal since 1940. The bastion of Singapore had fallen to the Japanese on 15 February 1942. By the end of that year, the Australians had been victorious on the Kokoda Trail, at Milne Bay, Buna, Gona and Sanananda, and at Lae. Early in 1943, Prime Minister Curtin and General Blamey realized that the direct threat to Australia was receding. The 9th Australian Division had returned to Australia and General Morshead had formed II Australian Corps consisting of the 7th and 9th Australian Divisions and a Brigade of the 6th Division. By the end of 1943, the Japanese were being heavily pursued in New Guinea. With the return of the AIF from the Middle East, Australia's participation in the Empire Air Training Scheme became the country's most significant contribution to the European war.

In November 1943, just as 455 Squadron was approaching its 1200th sortie and 6000 operational hours, the Squadron was told that it was to convert from the Hampden to the new Bristol Beaufighter. David Whishaw (then a Flying Officer in the Squadron) from Carrick, near Launceston in Tasmania remembers,

> There was a great stir when we learned we were to convert to Beaufighters and our future role was explained. Soon after, Jack Davenport arrived to command the Squadron together with Lloyd Wiggins and Colin Milson to take charge of 'A' and 'B' Flights.[45]

On 5 December 1943, the Commanding Officer, Wing Commander Bob Holmes, was posted out for repatriation to Australia and was succeeded by Wing Commander Jack Davenport DFC. Operational flying with Hampdens finished in December 1943 and the Squadron Operation Record Book records,

> 13 December 1943:
> Squadron comes out of line from Operations to commence rearmament program on Beaufighters.

A new chapter in the 455 Squadron story began.

2

First Beaufighter Operations

It was very exciting to fly. With the two wing mounted engines there was nothing in front of the cockpit to block the pilot's vision of the clouds and sky.

Peter Ilbery, former 455 Squadron Pilot, on the Beaufighter.

Brilliant leading down-sun on the Wing Leader's part in taking the convoy by surprise.

Surface Attack Report, 455 Squadron, 6 May 44.

The area between Ushant [Island, off Brest] and Borkum cannot now be a popular run for the German coast-wise sailor.

RAF Coastal Command Review, May 1944.

BEAUFIGHTERS began to arrive at the Leuchars station just before Christmas 1943. They arrived quickly - in two weeks, sixteen of these aircraft were with 455 Squadron. Conversion training started soon after and the crews were thrilled with the new machine. The Squadron Operations Record Book records, on 26 December 1943, 'Wing Commander Davenport flew Beaufighter T/455 for the first time and completed several successful circuits and bumps. The Australian Flag was hoisted for the first time outside Squadron Headquarters.' Pilot conversion from Hampdens to the Beaufighter could be an unceremonious affair as Jack Davenport recalled,

A Bristol test pilot talked to us about the plane and took me for a familiarization flight of twenty minutes - I stood behind him in the cockpit. When we got back to the airfield I got into the pilot's seat and the test pilot began to climb down from the plane. I asked him where he was going and he said, 'You don't expect me to go up with you, do you?' After that first flight I instructed most of the key Squadron personnel who in turn converted their own pilots and aircrew. Most of us were fairly experienced pilots and had taken the opportunity to fly different aircraft whenever we got the chance.[1]

All pilots seemed to enjoy flying the Beaufighter. They described it as tough, reliable and fast. Lloyd Wiggins, from Woodville, South Australia, was one of the Squadron's first Flight Commanders on Beaufighters. Before the War Wiggins had been an auctioneer and he joined the Air Force in Adelaide in 1940. He learned to fly in Western Australia, then travelled to England where he converted to Wellington bombers. Wiggins volunteered to join 38 Squadron RAF, which was converting for torpedo work in the Middle East. During the fighting for El Alamein in October 1942, when the RAF in the Middle East had been concentrating on halting Axis resupply to North Africa, a convoy was found heading towards Tobruk. A formation of three Wellingtons from 38 Squadron, led by Wiggins, attacked the convoy. They launched six torpedoes and sank a 5,800 ton ship; one Wellington was shot down and Wiggins' aircraft was also hit. For this action, Wiggins was awarded the DSO. He was posted back to Britain shortly after, with an operational tour completed, to instruct on torpedo work. Then he joined 455 Squadron at Leuchars where he converted to Beaufighters. He says, 'The Beaufighter was a beautiful aircraft, smooth as silk and with no vices. It was strongly built and withstood a lot of punishment. Having spent a lot of time flying Wellingtons with Pegasus engines the Beaufighter felt like a racer with its two powerful Hercules engines'.[2]

Peter Ilbery's experiences in learning to fly the Beaufighter were typical. He recalls,

> The conversion course was first to the Beaufort. It had dual controls and many of the characteristics of the Beaufighter but it was 'heavy'. In the conversion to Beaufighters one stood behind the instructor because there wasn't room for two in the cockpit. Everything happened so much faster in this aircraft. After a few hours tuition the instructor vacated the aircraft to the pupil. It was very exciting to fly. With the two wing-mounted engines there was nothing in front of the cockpit to block the pilot's vision of the clouds and sky.[3]

The Beaufighter was a big aircraft. Loaded, it weighed over eleven tons, compared to the Spitfire which weighed just over two and a half tons loaded. The pilot's cockpit was crowded with instruments and controls. The throttle, propellor and supercharger controls all had their separate gates. There were four levers on the left of the dashboard for normal and emergency hydraulics, and for undercarriage and flap control. The three trimming controls were on the right. Take-off checks were fairly simple: hatches, hydraulics, trim, mixture, air, pitch, fuel, flap, gills, gyro. The navigator would recite these checks to the pilot. Take-off was fairly straight forward, provided the pilot opened the throttles separately to maintain directional control until rudder power was felt. After that, the throttles could be taken straight through the gate while the Beaufighter accelerated before it floated off the ground.

The groundcrew had their own conversions to do. Flight Sergeant Jack 'Jeep' McKnight was an engine fitter with 455 Squadron and one of the

'The cockpit was crowded with instruments and controls': Cockpit of the Bristol Beaufighter.

original Squadron members, joining as an aircraft fitter in May 1941 at Williamtown. He had started work on Hampdens at Swinderby, moved with the Squadron to Leuchars and was there when the Beaufighters arrived. He says,

> We were sent in small groups to the Bristol factory to do a two week course. The Hercules engine on the new Beaufighter was a great engine. It had dual fourteen cylinder air-cooled engines with sleeve valves. This meant it had no tappets or valves like a conventional engine, but a sleeve inside the cylinder that would lift to expose inlet and exhaust ports. Because of its overall design the engine could take a lot of punishment and was very reliable. The plane could still be flown even with a couple of cylinders not working. Unfortunately the engine wasn't very easy to change; a complete engine change could take a whole day.[4]

The last Hampden bomber left 455 Squadron from Leuchars for disposal on 29 December 1943. The Squadron Operations Record Book notes, 'It is two years and five months since No. 455 Squadron received its first Hampden'.[5] That day, also, the first Squadron Beaufighter fatality occurred. Flying Officer Peter Gumbrell RAF was on a familiarization flight in an aircraft flown by a pilot from the New Zealand 489 Squadron. Spectators watched the Beaufighter crash in a low speed stall south of Leuchars and burst into flames. Both men were killed.[6]

Conversion to Beaufighters continued during January and February 1944.

Layout of the Bristol Beaufighter Cockpit

1. Supercharger control
2. Carburettor air-intake control (starboard)
3. Carburettor air-intake control (port)
4. Fuel cock control wheel (starboard tanks)
5. Fuel cock control wheel (port tanks)
6. Radio tuner
7. Airscrew speed control (port)
8. Airscrew speed control (starboard)
9. Mixture control
10. Throttle levers
11. Undercarriage control lever
12. Flap control
13. Flap position indicator
14. Hydraulic power lever
15. Hydraulic emergency selector lever
16. Oxygen delivery gauge
17. Oxygen supply gauge
18. Cockpit Illumination
19. Radio Altimeter
20. Undercarriage and tail wheel position indicators
21. Pneumatic system triple pressure gauge
22. Vacuum pump change over cock control
23. Clock
24. Blind approach visual indicator
25. Fire extinguisher buttons
26. Airspeed indicator
27. Altimeter
28. Vacuum gauge
29. Undercarriage and tail wheel indicator switch
30. Main magneto switches (port and starboard)
31. Rudder pedals
32. Control column
33. Oil temperature gauge (port) (obscured by control column)
34. Direction indicator (obscured by control yoke)
35. Artificial horizon
36. Reflector sight mounting
37. Rate-of-climb indicator
38. Firing button
39. Turn and bank indicator
40. Oil temperature gauge (starboard)
41. Fuel pressure gauge (port)
42. Fuel pressure gauge (starboard)
43. Radiator temperature indicator (port)
44. Radiator temperature indicator (starboard)
45. Boost gauge (port)
46. Boost gauge (starboard)
47. Air temperature gauge
48. RPM indicator (starboard)
49. RPM indicator (port)
50. Oil pressure gauge (port)
51. Oil pressure gauge (starboard)
52. Floodlight dimmer switch
53. Ventilator
54. Starter buttons (shielded) (port and starboard)
55. Intercommunication lamp and button (partially obscured by elevator trimming tab control wheel)
56. Elevator trimming tab control
57. Hydraulic hand pump
58. Compass
59. Rudder trimming tab control
60. 'Abandon aircraft' and pressure head heating switches

Air crews were busy with ground training, flying training and navigation exercises. The Squadron Operations Record Book reflects an optimism for an early return to operations,

> 30 January 1944:
>
> The rearming program is proceeding even better than expected and we hope to be fully operational by 28 Feb 44. The groundcrews have done a wonderful job keeping six Beaufighters serviceable for APC [air plane conversion] flying under the existing difficulties. We have 19 Beaufighters Mark X and one Oxford Mk II on our charge as at 31 Jan 44.[7]

At the pay parade on 21 January a competition was announced. A prize of one guinea would be awarded to the Squadron member who submitted the winning design for a Squadron crest. Later that year, in June, a rough copy of a crest showing a winged Viking helmet was sent to Royal Australian Air Force Overseas Headquarters in London. The Squadron Commanding Officer, Jack Davenport, explained the crest in his letter, 'The type of work carried out by this Squadron is akin to that of the Viking - short sharp swoops across the sea by a compact, well armed force...The spirit of the Australians is like that of the Vikings, adventurous and free.'[8] The motto chosen was 'Strike and Strike Swiftly'. That motto was already in use, so Davenport chose 'Strike and Strike Again'.[9]

By 23 February, all crews had finished conversion training. On 1 March the Squadron was operational again, designated a fighter-reconnaissance squadron. Along with 489 Squadron RNZAF, they would be known as the Leuchars Wing and would work together on anti-shipping strikes off the Norwegian coast.

489 Squadron RNZAF had formed in August 1941, at Leuchars. Crews had begun to train on Beauforts but when the situation in the Mediterranean became serious, these aircraft were transferred at short notice to attack Rommel's supply ships and 489 Squadron was re-equipped with Hampdens. Training continued until April 1942 when the Squadron moved to the south of England to begin their operational role. The Squadron flew anti-submarine patrols, far out into the Atlantic and deep into the Bay of Biscay. 489 Squadron returned to Leuchars in July 1942, joining 455 Squadron, and there they began their long career together attacking German shipping. 489 Squadron converted to Beaufighters with 455 Squadron at the end of 1943.

March 1944 was an important time for the Allies. The previous year had been a year of almost unbroken success - the Allies were dominant in the air and defeats in Russia and the Mediterranean had cost Hitler more than a million men killed or captured. Although Nazi Germany was still far from defeated, Admiral Doenitz, later Commander-in-Chief of the German Navy, was faced with his 'most difficult decision of the war'. He wrote later,

> I finally came to the bitter conclusion that we had no option but to fight on. The U-boat Arm could not stand aside and watch the onslaught, of which it had hitherto borne the brunt, now fall in all its fury as an additional burden on the other fighting services and the civilian population.[10]

455 Squadron's first Beaufighter strike came soon after they were declared operational at Leuchars. They were led by Squadron Leader Colin Milson DFC, an experienced and highly respected operational pilot. Colin Milson was a direct descendant of James Milson, who had been a friend of Joseph Banks and a free settler to Australia, arriving from Lincolnshire in 1806. James Milson was given land in Sydney by Governor King, at a place which became known as Milson's Point. Milson's maternal grandfather, George Morgan-Reade, was one of the founders of QANTAS. Morgan-Reade's son, Colin Morgan-Reade had been killed at Gallipoli and Colin Milson was named after him.[11] After leaving school, young Colin Milson went to his family's property 'Springvale' at Winton, Queensland, for six months before working at Triangle, New South Wales, as a jackeroo. Soon after the War started Colin applied to join the AIF, but changed his mind a few days later and applied for the RAAF. He was refused at first because he lacked flying experience (only candidates with over 100 hours flying were being considered) but was accepted in 1940. He was then twenty years old.

Colin Milson finished his flying training at an Operational Training Unit in England and was posted to 39 Squadron RAF, already in action in the Middle East, flying Bristol Beaufort torpedo bombers. 39 Squadron was sent to North Africa, to an airstrip near Cairo. From there they were pushed back to the Suez Canal zone when the Germans drove through to Alamein. 39 Squadron was also detached to Malta several times to help disrupt the flow of supplies to the Germans in North Africa. On 6 September 1942, Milson lead a force of twelve Beauforts from Malta to attack a south-bound convoy off Cape San Maria Di Leuca, near the 'heel' of Italy. They attacked the convoy against escorting destroyers and fighters. On another detachment to Malta in February 1943, Milson led his flight in an attack on a large tanker south-west of Marsala in Sicily. Although the tanker was escorted by three destroyers and a force of fighters it was destroyed.[12] For these two attacks and his other operations with 39 Squadron, Milson, then a Squadron Leader, was awarded the Distinguished Flying Cross.

After almost three years away from home, Milson was keen for a break. He returned to London on leave in May 1943 and in late June was sent to the Air Ministry. After Malta he found London crowded and expensive, and he wrote to his father, 'Would much rather be back on a Squadron, but would be rather foolish to chuck up my present work, although I would be much happier on flying'.[13] After some three months in London Milson converted to Beaufighters and was sent to Leuchars briefly, with 455 Squadron. He was then sent on an unofficial attachment to 144 Squadron RAF, flying Beaufighters from the Coastal Command Station at Wick in Scotland. Milson finished his first operational tour with Coastal Command in December 1943, when he joined 455 Squadron as one of Davenport's Flight Commanders.

On the afternoon of 6 March 1944, Milson led eight 455 Squadron Beaufighters providing anti-flak protection to four 489 Squadron RNZAF Beaufighters on an anti-shipping patrol off the Norwegian Coast. With Milson on that first Beaufighter strike was Flight Sergeant John 'Tiger' Payne, on his first operational sortie. Payne was a salesman from northern New

South Wales, near Lismore. He had joined 455 Squadron on 29 November 1943.

The twelve Beaufighters of the Leuchars Wing took off at 3 pm and formed up with the 489 Squadron torpedo aircraft in the centre and four anti-flak aircraft on each wing. Off the southern coast of Norway, in the vicinity of Egero Light, they found a sixteen vessel convoy heading north; six of them merchant ships. The convoy was protected by nine escort vessels and an air escort of four single-engine Messerschmitt Bf109s and a Blohm and Voss flying boat. The Beaufighters attacked from the seaward side of the convoy, in the face of flak from the escorts and shore batteries. With their 20 millimeter cannon, 455 Squadron scored numerous hits on the decks and superstructures of the merchant vessels and escorts. Three of the 489 Squadron Beaufighters made good attacks with their torpedoes, but the fourth was intercepted by a Messerschmitt before it could get into a dropping position and had to find cover in the clouds. The 455 Squadron Beaufighter flown by Flight Sergeant Payne was also chased and fired on by a Messerschmitt during his run-in, but was not hit, and he escaped into cloud. Milson had his Beaufighter holed by flak but was able to return to base. One merchant vessel was seen to explode and was later confirmed to have sunk.[14] The Wing returned to Leuchars and landed at 6.40 pm, after some three and a half hours flying. For leading the Australian Beaufighter pilots on their first operational sortie and for carrying out a successful attack against heavy opposition, Colin Milson was recommended for a Bar to his Distinguished Flying Cross.

During the rest of March 1944 the Leuchars Wing suffered the frustrations of poor weather off the coast of Scotland - the weather was often unsuitable for any flying at all. They managed to get off the ground for five more patrols from Leuchars but bad light, poor weather and lack of sightings of enemy shipping gave no further results. For that month, Coastal Command recorded four major strikes off the Dutch Coast and three off the Norwegian Coast, including the Leuchars Wing attack on 6 March.

With Operation "Overlord" approaching, Leuchars was too far north for the strike wing. Both 455 and 489 squadrons were ordered south to RAF Station Langham. The Langham field was several kilometres from the sea on the north coast of Norfolk, near the towns of Cley-next-the-Sea, Blakeney and Stiffkey, and nearly forty kilometres north-west of Norwich. It was a good location for a Coastal Command airfield; close to the sea, on flat ground and with clear approaches. 455 Squadron was taken off the fighting strength of Coastal Command's 18 Group on 2 April 1944 and withdrawn from operations.

On 6 April, the 455 Squadron advance party of thirteen left Leuchars on a passenger train to Langham. They arrived next afternoon and by 8 April, the new Squadron Orderly Room was functioning at Langham. Several days later the main party left Leuchars by special train, under the command of Squadron Leader Lloyd Wiggins. They arrived in Norfolk, at Thursford Railway Station, on the afternoon of 12 April. Finally the air party left Leuchars and upon their arrival at Langham on 13 April they 'shot up the 'drome in fine style'.[15] The runway at Langham had only just been resurfaced when the Squadron aircraft began to arrive. Jack McKnight recalled the groundcrews' horror at the arrival of the aircraft,

> As the Squadron aircraft arrived we found that the runway had recently been sprayed with tar. That, combined with the unusually warm weather meant that tar was splashed on the hydraulic piston arms of the undercarriage. These had to be kept spotlessly clean to avoid damage to the seals on the the hydraulic cylinders. The airframe fitters were nearly in tears when they saw this because they'd have to laboriously clean off the tar with rags and kerosene.[16]

Accommodation at Langham was in Nissen style huts, with some twenty to thirty men in each. The huts were heated by a cast iron fuel stove that could burn any type of fuel, though coal was mainly used. There was a separate building for ablutions and washing. The Nissen huts, messes, control tower and operations rooms were well separated as protection against German air attack. 455's move to Langham was finished by mid-April. With 489 Squadron RNZAF they formed what became known as the Langham Wing or, to some, the Anzac Wing.

The Station Commander at Langham was Group Captain Arthur Clouston. Clouston was a greatly respected commander and a remarkable airman. Born in New Zealand, he was eighteen years old when Charles Kingsford-Smith flew across the Pacific from America to Australia in his 'Southern Cross'. Clouston saw Kingsford-Smith's aircraft in Christchurch and he was determined to fly. Clouston travelled to England, joined the RAF, and then became one of the first civilian test pilots at the Royal Aircraft Establishment, Farnborough. As part of his work Clouston experimented with flying into wires to determine the effect on an aircraft when it flew into the wires of a barrage balloon. In 1936 he was nearly killed in the Schlesinger air race from England to Johannesburg, and he barely survived flying across the Swiss Alps in the 1937 air race from New York to Paris. Late in 1937 he broke Amy Johnson's London to Cape record and he was awarded the Britannia Trophy for the best performance in the British Empire in the air for that year. In 1938 Clouston broke all existing records for the London to Australia flight, and London to New Zealand and back. The England – Australia – England record was now set at 10 days and 21 hours, during which Clouston and his co-pilot had only a total of sixteen hours sleep.

When War broke out, Clouston was back in the RAF as a Squadron Leader, still test-flying at Farnborough. He helped develop a searchlight to be installed in the nose of a twin-engined American bomber (known as the Helmore Light after the Project leader Air Commodore Helmore, a retired RAF officer). Clouston also flew for Helmore in a project to control high powered launches from the air, and he asked the Commander-in-Chief of Coastal Command, Air Marshal Sir John Slessor, to be able to fill the launches with high explosive to use against the German battleships *Scharnhorst* and *Gneisenau* (then in Brest harbour). Slessor refused. Clouston then helped Wing Commander Leigh at Coastal Command Headquarters to fit a searchlight to a Wellington bomber to illuminate German submarines at night. The project was very successful. Finally, and at his own request, Clouston was transferred to command 224 Squadron RAF, flying anti-submarine Liberators fitted with the 'Leigh Light'. He was awarded the Air Force Cross in 1938 and the Bar in 1942. Early in 1944, Clouston was promoted Group Captain and sent to command the new Coastal Command station opening at Langham. According to a New Zealand Beaufighter pilot,

He was as straight as a die, sincere, knowledgeable and humane. He couldn't be bothered with the typical RAF bull, such as parades and saluting but he certainly knew how to join in with the crews under strain letting off steam, by throwing a rip-snorter party. He did have a very unusual weakness which occasionally burst to the surface when the party had reached its peak, with most people well oiled up. This required two ingredients - an old off-key piano and a large fireman's axe. The first was always found at one end of the mess and the second in the glass cabinet on the wall just outside the foyer. Arthur Clouston held the record for piano demolishing at Langham, about 10 seconds! I don't know how much fuel he needed to reach that stage but it surely must have been considerable as his lifting arm acted as a mighty fast primer. Next morning he'd be there for breakfast even if most of us were 'missing, believed incapable'. He was truly a remarkable man.[17]

The Australian and New Zealand squadrons were now part of Coastal Command's 16 Group which had its headquarters near Chatham, south of London. Air Marshal Sir Sholto Douglas had just succeeded Air Marshal Slessor as Commander-in-Chief Coastal Command. Douglas says,

In April, a couple of months before the invasion was due to take place, I issued all my Group Commanders a directive which set out in full detail the tasks which each of the Groups would be expected to perform. So far as we could see the main threats to our cross-Channel shipping would be from U-boats coming from (bases) in the Biscay ports, from destroyers and torpedo boats from Brest and Cherbourg, from more destroyers and torpedo...boats coming down from Dutch ports, as well as from Le Havre and Boulogne.[18]

16 Group's orders were to strangle enemy shipping routes in the North Sea and prevent German motor torpedo boat attacks against British convoys. 455 Squadron received orders from Group Headquarters on 14 April - the Squadron was now to be used in the bombing role against German shipping. Over the next few days the crews went to lectures on anti-shipping tactics, aircraft compasses were swung and crews practised bombing at a nearby range.

Operations from Langham began almost immediately. On 19 April, eight Beaufighters left Langham separately on shipping reconnaissance patrols along the Dutch Coast. Squadron Leader Lloyd Wiggins and Flying Officer Bill Barbour, a bank clerk from Chinchilla, Queensland, both sighted large convoys. Beaufighters from the strike wing at nearby North Coates, south-east of Grimsby, attacked this shipping with good results. 455 Squadron was congratulated for its work by the Air Officer Commanding 16 Group.

At the end of April 1944, Coastal Command noted that 'Off the coasts of

Norway and the Frisian Islands, in the North Sea, the Channel and the Bay, German shipping has been energetically harried during the month. The Beaufighter Wings have again done well.'[19] During April, Coastal had flown 720 anti-shipping sorties. The Beaufighter wing at Wick in the north of Scotland had attacked a convoy near Stadlandet. Off the Dutch Coast, the well-practised North Coates Beaufighter wing, operating three Beaufighter squadrons since late in 1942, had made five major shipping attacks.

On 30 April 1944, at the new Coastal Command station at Langham, the Commanding Officer 455 Squadron, Wing Commander Jack Davenport, had under his command twenty-one Beaufighters, one Oxford Mk II, and 219 all ranks including 77 aircrew.

During April 1944, preparations for the invasion of Europe continued.

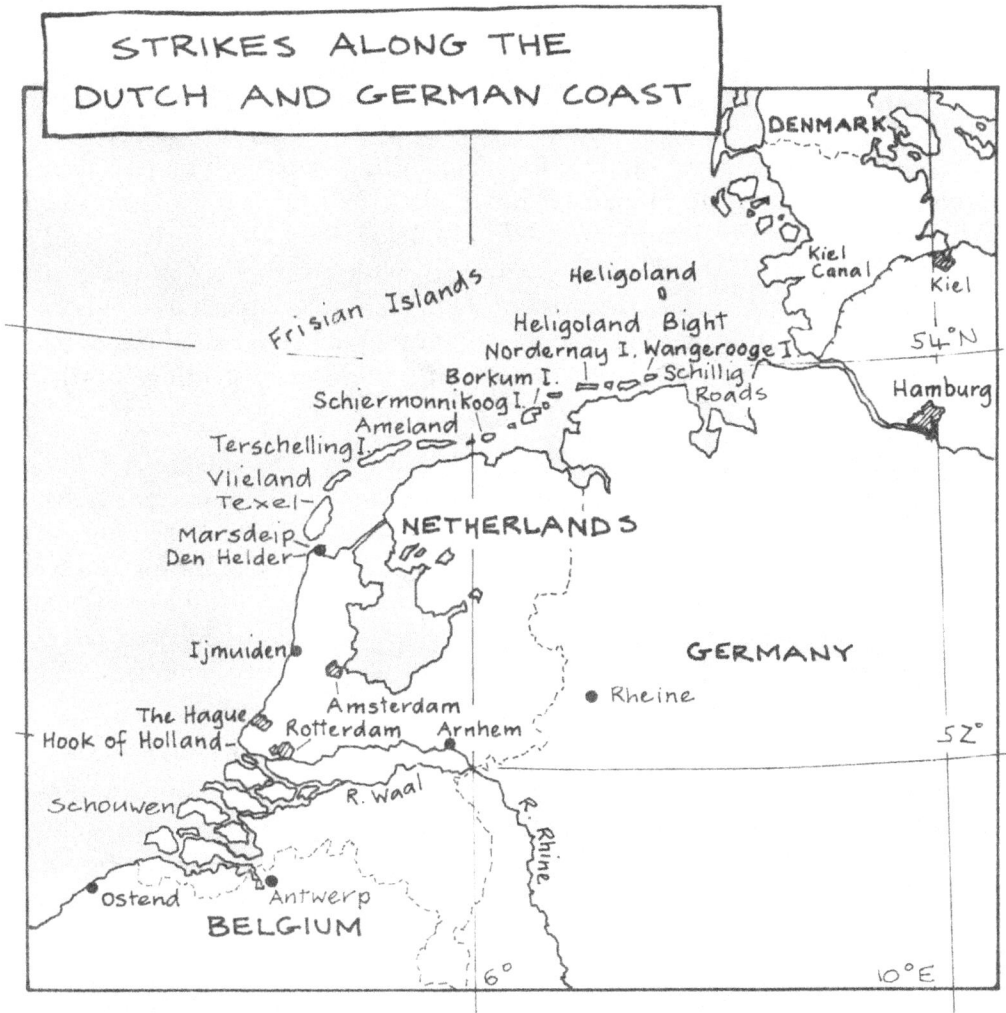

STRIKES ALONG THE DUTCH AND GERMAN COAST

American bombers began attacks on German shore batteries along the Normandy coast, and they also bombed anti-aircraft batteries between Dover and Dunkirk to maintain the deception of a landing in the Pas de Calais. To maintain the deception, it was laid down that for every gun battery bombed in the actual assault area, two batteries had to be bombed elsewhere. Ultra signals decrypts were giving British Intelligence a good picture of the disposition of German forces in north-western France and it seemed that the scale of German strength needed to postpone or cancel D-Day would not be reached. "Overlord" would go ahead.

There was a bad air crash near Langham late on 1 May. Pilot Officer Neville Wilson had taken off from Langham and his port engine failed almost immediately. He was at eight hundred feet, with wheels and flaps up. Wilson flew for some time on his good engine, then made for a nearby airfield at Little Snoring. He made an even, level approach to the runway, but lost height just short of the field and sank into a wood (the investigation concluded that he had either throttled back or the starboard engine had overheated). Wilson's navigator, Flight Sergeant Ted Holmes RAF, fought his way out of his cupola and made three attempts to reach Wilson, but the pilot's canopy was covered in burning trees. Wilson died and Holmes suffered facial burns.

Very early on 4 May 44, the Langham Wing took off for an armed reconnaissance along the French Channel coast. Led by Wing Commander Jack Davenport, thirty Beaufighters left at 4.45 am and over an hour and a half later they approached the Normandy coast with orders were to attack any shipping encountered inside their patrol area. They flew on at three hundred feet, in perfect conditions, and just after 6 am they received a radio message, 'Possibility of shipping 5 miles north of Cherbourg steering north'.[20] Davenport recalls,

> I remember we were really looking for trouble that day. Towards the end of our search area we had found nothing, then I saw a couple of ships to our starboard flank: they appeared to be the size of German E-boats and I detached my right flank under Flight Lieutenant Pilcher to attack. I took the rest of the formation to the end of the patrol area then headed back towards the ships we'd seen earlier. I caught up with my right flank and was just about to attack when I saw the red ensign on one of the ships. The ships had opened fire and several planes had attacked with cannon before we were able to call off the attack.[21]

The ships were Royal Navy motor torpedo boats and they had actually fired on the Beaufighters. They did not fire the 'colours of the day' until a number of the Beaufighters had attacked and the ships realized what was happening. All the boats sank after the attack and Jack Davenport sent the force back to Langham and stayed with the survivors to make sure a good fix was given on their position. He says,

> All hell broke loose over this and I expected to be court martialled. Investigations went on for three to four days and my only defence was that we'd been told to attack any shipping south of a particular line and that's what we did. After several more days I got a telegram from the Admiral in Portsmouth congratulating the Squadron and remarking that three Royal Navy ships had been sunk with forty-two rounds of 20 millimeter cannon.[22]

In 1944, Germany had been forced to make increasing use of coastal convoys to overcome the shortcomings of their internal transportation system, disrupted by Bomber Command in the lead-up to the D-Day invasion. German merchant shipping was well organized and well-defended. The activities of the German merchant marine were under a single authority, the *Reichs Kommisar für Schiffahrt*. In 1941 its head, Karl Kaufmann, assumed control of over 500 vessels in North West Europe, totalling 946,598 tons. His cargo vessels ranged from one thousand to ten thousand tons. All ships were armed, the weapons being served by well-trained naval gunners. The merchant vessels were protected by the *Kriegsmarine* (German Navy) in the ratio of two or three escort vessels to each merchant ship.

A large proportion of the *Kriegsmarine* was concerned solely with defensive duties, such as minesweeping, convoy escort, defensive mine laying and harbour defence. The most common escorts for the merchant convoys were converted trawlers, usually about five hundred tons. They were armed with flak guns of all calibres and were capable in many cases of up to twelve knots. The Germans called these *Vorposterboote* and the RAF called them 'trawler-type auxiliaries'. To aircrew they were 'flak ships'. A typical armament would be one 88 millimetre gun, two or three 20 millimetre guns and a number of machine guns. They frequently flew balloons and some had a form of flame-thrower at the mast head.

There was a larger and more feared version of the flak ships called the *Sperrbrecher,* meaning 'barrier-breaker', whose primary function was mine-sweeping. These were converted merchant ships of between 1,500 and 8,000 tons, specially reinforced for exploding mines and packed with flak defences. They were heavily armed with 88 millimetre, 37 millimetre and 20 millimetre guns. It was estimated that Germany had fifty or sixty of these specialised ships. They were identified by 'dazzle' painting, cut down masts and many gun positions. Coastal Command Intelligence summaries also noted that these features, 'usually leaves little doubt as to the identity of these vessels' and, 'In addition there are one or two swastikas painted on their decks, for those who can get near enough, to note'. The strike squadrons frequently did.

There were also purpose built mine-sweepers and escorts that usually sailed just ahead of the convoys. These were *Minensuchboote* or M-Class minesweepers. It was estimated that Germany had 120 to 150 of these ships. These boats could reach speeds of seventeen knots and when escorting they frequently flew balloons. Smaller and more agile minesweepers sometimes escorted the convoys. They were *Raumboote* or R-boats. Averaging between 115 and 190 tons they were a dangerous prospect for aircrew with their one or two 37 millimetre guns and two 20 millimetre guns. Some types had multiple 20 millimetre guns.

Coastal Command kept a close watch on German convoy habits, and re-gularly distributed a Secret bulletin to stations. In March 1944, the bulletin noted some change in convoy habits – 'First, lengthening daylight has seen a later time of departure of Northbound convoys from the Hook [of Holland] which was to be expected.' The importance of the harbour at the Dutch port of Den Helder was also noted. 'Further, it appears that the north and westbound shipping have adopted the practice of sheltering in the Helder anchorage during the day: it will be remembered that up to February it was the general rule for the former to sail straight through to the Elbe, whereas in March, probably only one convoy did so.'[23]

The Commanders at Langham in May 1944: The Commanding Officer of 455 Squadron, Wing Commander Jack Davenport DSO, DFC (centre) with his two flight commanders, Officer Commanding A Flight, Squadron Leader Lloyd Wiggins DSO (left), and Officer Commanding B Flight, Squadron Leader Colin Milson DFC. All three were on their second operational tours.

During the afternoon of 6 May, a reconnaissance patrol from 489 Squadron reported a convoy of merchant ships and escorts off Borkum, heading south-west. Borkum is a small island off the German coast at the eastern end of the Dutch Frisian Islands and on the main shipping lane with the Dutch port of Rotterdam, then occupied by the Germans. Coastal Command ordered an immediate strike from Langham. Just two and a half hours later, at 6.15 pm, a formation of eighteen Beaufighters closed in on the Dutch Frisian Islands. The Langham Wing had sent twelve Beaufighters from the Australian Squadron and six from the New Zealand Squadron. The New Zealand aircraft carried torpedoes and the Australian Squadron's job was to suppress the flak – three anti-flak planes for every torpedo-carrying aircraft. Starting their patrol from north of Terschelling Island, the Beaufighters swept up the main coastal shipping lane towards Borkum. They were down close to the sea and in formation – an anti-flak section on either side of the torpedo carriers.

It was bright, with some sea haze and there were white-caps on the waves from the moderate wind. When they saw the target, some twenty-five kilometres north-east of Borkum, the formation wheeled round to the south-west away from the shore flak guns, and climbed to eight hundred feet. With the sun now behind them, they attacked the convoy head-on.

This convoy was heavily defended. There were six merchant ships and twelve escorts, and the escort vessels surrounding the merchant ships flew balloons on heavy wires. When they saw the Beaufighters they opened up with a flak barrage; it began a furious exchange.

The anti-flak Beaufighters came in first in a shallow dive. Six Australian pilots took on the leading escorts, strafing them with cannon. Several of the pilots went on, flying the length of the convoy to attack the rear. Flight Sergeant Jack Costello of Homebush, New South Wales, attacked a leading escort and a merchant ship, then strafed an R-boat at the rear of the convoy. He expended 750 of the 960 rounds of cannon that his plane carried, despite having a cannon stoppage during his attack. A second wave of Beaufighters flown by the New Zealanders came in low and level, dropping torpedoes in front of the merchant ships. The 455 Squadron Operations Record Book records the result,

> The formation turned into attack out of the sun and the convoy was apparently taken by surprise. The aircraft were subject to intense flak at first, but very meagre flak at the end of the encounter. All aircraft observed many hits all over their targets. Number 489 Squadron released all their torpedoes at the convoy.[24]

A three thousand ton merchant ship was torpedoed and seen to be on fire low in the water while another ship of two to three thousand tons was also left on fire.[25] Eight other vessels were damaged. The station commander, Group Captain Arthur Clousten noted later in the attack report, 'Brilliant leading down-sun on the Wing leader's part in taking the convoy by surprise after a 69 mile patrol in sight of the Dutch Coast'. This same convoy was attacked again by twenty-three Beaufighters from the nearby North Coates strike wing just half an hour later.

455 Squadron lost their first crew from Langham that day. Before the attack, the Beaufighter piloted by Flight Lieutenant Byron 'Pip' Atkinson RAF, had been lagging behind the port section anti-flak group. Flying Officer Neil Smith from Brighton, Victoria, was in formation and saw Atkinson's plane dropping behind. He says,

> I can recall looking back on the port side and seeing this straggling aircraft on its own : we had always been told to keep in tight formation mainly for protection against possible enemy fighter attacks but also for concentrated attack by us as anti-flak aircraft. It was quite some time before I realized that that aircraft must have been 'Pip' Atkinson's.[26]

The aircraft failed to return to Langham and the pilot and his navigator, Flight Sergeant 'Shorty' Whitburn were posted as missing. One crew reported that they had seen a splash and some smoke some three kilometres to the starboard of the convoy. Byron Atkinson had joined the Squadron on

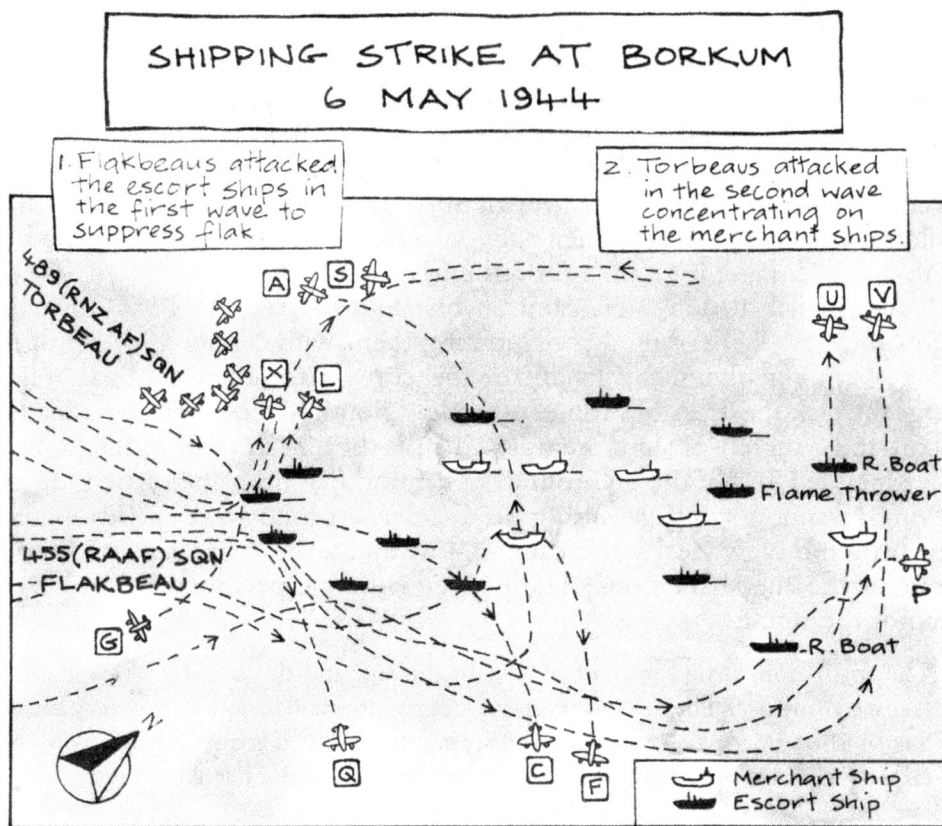

SHIPPING STRIKE AT BORKUM 6 MAY 1944

1. Flakbeaus attacked the escort ships in the first wave to suppress flak.

2. Torbeaus attacked in the second wave concentrating on the merchant ships.

489 (RNZAF) SQN TORBEAU

455 (RAAF) SQN FLAKBEAU

R.Boat
Flame Thrower
R.Boat

⌣ Merchant Ship
⚓ Escort Ship

455 Squadron provided anti-flak protection for the torpedo Beaufighters of 489 Squadron RNZAF in an anti-shipping strike off the German Coast on 6 May 1944. 455 Squadron aircraft and crews were:

A	F/Sgt R. Walker RAF (pilot) F/Sgt T. Rabbitts RAF (navigator)	P	F/O D. Whishaw (pilot) F/O J. Belfield RAF (navigator)
C	S/Ldr A.L. Wiggins DSO (pilot) F/Sgt R.F. Day RAF (navigator)	Q	F/O N.R. Smith (pilot) F/O F. Macintyre (navigator)
F	F/Sgt G.E. Batchelor RAF (pilot) F/Sgt H.R. Morris RAF (navigator)	S	S/Ldr C.G Milson DFC (pilot) F/O M.F. Southgate RAF (navigator)
G	W/O N.P. Turner (pilot) F/Sgt G.F. Hammond (navigator)	U	F/Sgt J. Costello (pilot) F/Sgt H.A. Hufford RAF (navigator)
L	F/Sgt J.C Payne (pilot) F/Sgt J. Rennie RAF (navigator)	V	F/O I.H. Masson RAF (pilot) F/Sgt E.H. Knight RAF (navigator)
M	F/Lt B. Atkinson RAF (pilot) F/Sgt J.A. Whitburn (navigator)	X	F/Lt J.M. Pilcher (pilot) F/O S.E. Drinkwater RAF (navigator)

5 April 1943 and had completed nineteen operational sorties. Whitburn had joined the Squadron only a month before and this had been his first operation.

On 9 May, nine RAF navigators were posted out of the Squadron. The Squadron Operations Book records that, 'These RAF navigators were posted in accordance with a policy which is to make No 455 (RAAF) Squadron one hundred percent 'Aussie' aircrew!' That policy would take time to implement. By the end of May 1944, there were seventy-two aircrew in the Squadron, but only fifty-eight per cent were Australians. However, the ratio of RAAF to RAF would grow steadily for the rest of the War until, on its disbandment, the Squadron had only three aircrew who were not Australians.

Air Marshal Sir Sholto Douglas visited Langham on 12 May. The Commander-in-Chief of Coastal Command spoke to all the crews on the importance of their role in the success of the forthcoming invasion of Europe. Two days later, 14 May, the Squadron was on operations again. Wing Commander Jack Davenport led eighteen Beaufighters and an escort of Mustang fighters from Coltishall (north of Norwich), up the main Dutch coastal convoy route as far as Nordernay Island in the Frisian group. A convoy that had been reported wasn't sighted so they turned around to fly south-west again on a reciprocal course. Jack Davenport had six torpedo Beaufighters from 489 Squadron as his main weapon. He stationed twelve of his own Squadron Beaufighters in two echelons on the starboard side for flak protection and six anti-flak Beaufighters from 489 Squadron as escorts on the port side. It was early afternoon but rain, low cloud and sea haze made visibility poor – crews could see less than three kilometres.

Some twenty-five kilometres north of Ameland, and less than two kilometres off their port wing, they found a convoy. There were four merchant ships heading up the Dutch coast, ringed by sixteen escorts. This was heavy protection. Surprise was the attackers' ally, so Davenport wheeled his Squadron straight in. Being so close to the convoy there was little time to manoeuvre as they swept in from the north.

Three 455 Squadron anti-flak Beaufighter pilots converged on four minesweepers at the front of the convoy. Flying Officer Bob McColl from Koorawatha, New South Wales, attacked two of the leading minesweepers in his path, firing four hundred rounds of cannon. Jack Davenport and Flying Officer Ian Masson RAF attacked in a dive, pushing the throttles through to the stops, 'going full bore' across the back of the convoy, shooting at an escort each and then attacking merchant ships in their paths. Masson's was a difficult attack. Rain made it hard to see out of the front windscreen, the flak was heavy, and he took a direct hit on the top of his port engine. His cockpit was riddled with shrapnel, but Masson still managed to get away 460 rounds of cannon at the convoy.

The flak had been heavy at first but died down. This was what the Australians had been hoping to achieve; it gave the torpedo aircraft a level, steady run-in. The attack left one merchant vessel torpedoed and sunk and a second set alight. An armed minesweeper was also sunk. Five 455 Squadron Beaufighters returned to Langham with cannon, bullet or shrapnel holes. One aircraft from 489 Squadron crashed into the sea with the loss of the crew. When Ian Masson returned to base he found that his landing hydraulics had been shot away and he brought off another good belly landing; he had belly-landed a Hampden nine months earlier. Davenport noted on his attack report, 'Continuous rain made sighting difficult and a windscreen protection against rain is essential. Pilots made good attacks under difficult conditions and in the very little time available.'[27] For his work on this operation and earlier sorties, Masson was recommended for the Immediate Award of the Distinguished Flying Cross. Davenport wrote on the recommendation,

> Undeterred, in his usual brilliant manner, he effected a masterly 'belly-landing', causing very little damage to the aircraft and without injuring his crew. The above is typical of Flying Officer Masson. Energetic determination in the face of very heavy opposition, and a keenness to meet the enemy, is the keynote of his work, which is a brilliant example and inspiration to all.[28]

At the time of this award, Masson had completed twenty-seven operational sorties, involving 187 hours of operational flying.

In the weeks before D-Day, the Anzac Wing at Langham kept up their reconnaissance patrols along the shipping lanes. They also patrolled along the French and Dutch coasts looking for German torpedo boats and other light naval craft operating from bases at Ijmuiden on the Dutch coast and Cherbourg on the French coast. It was planned that on the eve of D-Day, Coastal Command and the Fleet Air Arm would cooperate with surface vessels of the Allied navies in a wide and complicated pattern of patrols to seal both the eastern and western approaches to the Channel. Coastal Command records show that, in May 1944, nearly seven thousand hours were flown on operations. Eighty-five attacks were made in the area between Ushant Island off Brest, and Borkum. The Coastal Command Review noted that this area, 'cannot now be a popular run for the German coastwise sailor'.[29]

Strike Wing tactics continued to evolve. Originally, Coastal worked on the principle that 'up to twelve aircraft should carry torpedoes, depending on the composition of the convoy, and that three anti-flak aircraft were necessary to deal effectively with each escorting vessel.'[30] But now the Germans were increasing the protection of their convoys. Jack Davenport recalled that, 'It was not uncommon for one merchant ship to be surrounded by up to fourteen escorts acting as shields. That was when we started the 'big wing' tactics.'[31]

The combined wings in Coastal Command were reminiscent of Air Vice Marshal Leigh-Mallory's 'big wing' tactics proposed for 12 Group in Fighter Command during the Battle of Britain. On 31 May 1944, the Langham Wing and the nearby North Coates Wing joined for a trial of combined wing tactics. The two wings, Langham and North Coates, had each been using different strike techniques. The Australians and New Zealanders in the Langham Wing were used to flying very low across the water on the way to their target, gaining height just before the attack. 455 Squadron had always been a low-level squadron, even back in Bomber Command. They were also used to flying their anti-flak aircraft on the flanks of the torpedo aircraft, separating just before an attack. In contrast, the North Coates Wing usually flew their anti-flak aircraft at about two thousand feet, with their torpedo aircraft below and about two thousand metres behind. Now, on this combined wing exercise, the North Coates Wing led, followed by the Langham Wing about one kilometre behind. North Coates went into the target first.

After the exercise, Jack Davenport wrote a report to his Station Commander at Langham. He suggested that, on all combined wing strikes, the Langham Wing should lead. He argued that, if the North Coates Wing were to lead a strike, the element of surprise could be lost because North Coates flew higher. He also said that, because Langham flew their torpedo carrying aircraft close to the anti-flak aircraft, the second wave from North Coates would be able to follow on close behind, 'flooding the defences'.[32] These suggestions would be put into practice, and would prove decisive in the busy months ahead.

During May 1944, 455 Squadron had logged 330 hours of operational flying, and over 430 hours in training. This reflected the strong emphasis given to training throughout Coastal Command, an emphasis which had been expressed when the Senior Air Staff Officer at Coastal Command, Air Vice Marshal Ellwood, had visited Leuchars late in 1943. He talked to aircrew and followed with a personal letter to Jack Davenport. In this letter, Ellwood likened air-crew training to practice for a professional. His analogy was that a professional cricketer, or footballer, or musician, has to practice for hours everyday to maintain the high standard of skills that are needed in each profession. Ellwood said that his aim was to produce professional aircrew, and he appealed to Davenport to look to his Squadron's training programme as a means of ensuring that every man received the practice he most needed, 'and plenty of it'.[33]

The Langham Wing patrolled the Dutch and French coasts on the best days leading up to D-Day, but found no enemy shipping. On a night navigation exercise on 2 June, Flying Officer Bob McColl's starboard engine blew - one of the cylinders crashed through the engine cowling and was left, according to the Squadron Record Book, 'dangling in the slipstream'. He returned to Langham, flying 240 kilometres on his port engine, landing safely. The station commander, Group Captain Arthur Clouston wrote in McColl's log book, 'A particularly brilliant piece of Airmanship McColl'. From Arthur Clouston, this was high praise.

In the months leading up to D-Day, 455 Squadron had flown, on average, six anti-shipping sorties per month. Now, the pace was about to quicken.

3

D – DAY, 'the air must hold the ring'

Today is D-Day – the Allies land in France

455 Squadron Operations Record Book, 6 June 1944

The enemy...continued to increase the escorts to his precious merchant ships. Our counter-measure was to send a combined wing strike.

Coastal Command Review, on the evolution of strike wing tactics

I gained a little height and the tracer bullets followed me. I flew lower again and found that it took the gunners several seconds to adjust their aim. I had to repeat this several times before we were out of range.

Jack Cox, Beaufighter pilot 455 Squadron, on the shipping strike 29 June 1944

'WE shall be back'. Winston Churchill said this to the French in June 1940, just after the evacuation from Dunkirk. Now, after years of doubt, disappointment and prolonged debate at many conferences, the Allies were to go back into Europe. An invasion of Western Europe was to be the supreme operation of 1944, and US General Dwight Eisenhower was appointed Supreme Commander. The Combined Chiefs of Staff directed him, 'You will enter the continent of Europe and, in conjunction with the other allied nations, undertake operations aimed at the heart of Germany and the destruction of her armed forces.'

This was Operation "Overlord". The plan was to employ five army divisions in an initial landing on the Normandy coast. The amount of detailed work required for the whole undertaking was formidable and the planning intensified as the final invasion date approached. The final "Overlord" conference was held on 15 May, attended by General Montgomery, Winston

Churchill and King George VI. General Montgomery told the audience, 'We must blast our way ashore and get a good lodgement before the enemy can bring sufficient reserves up to turn us out...We must gain space rapidly and peg out claims well inland. And while we are engaged in doing this the air must hold the ring...'

Coastal was working hard. Its strike wings had all been moved south to bases in 16 and 19 Group areas to block German minor warships attempting to attack either flank of the "Overlord" invasion convoys. At the same time, it was protecting shipping in the Atlantic sea lanes and it continued to harass German shipping off the coast of Norway and the Frisian islands. Because of the frequent air patrols over these waters, German ships seldom sailed by day.

455 Squadron now had two roles. First, it provided anti-flak protection on wing strikes with the New Zealanders. For this it used the Beaufighter's 20 millimetre cannons to suppress flak while the torpedo-carrying Beaufighters attacked the merchant ships. Second, it flew armed reconnaissance sorties, working singly and in small groups, carrying 500 pound and 250 pound bombs to use against smaller German boats.

Throughout most of May 1944, the weather for the Allied invasion had been almost ideal. But the first days of June brought a gradual deterioration and with it a series of dramatic conferences at Southwick House near Portsmouth. Eisenhower and his commanders were meeting daily to finalise last minute preparations for "Overlord" and to receive the weather forecasts upon which the final decision for the date of the assault depended. D-Day had been provisionally fixed for 5 June and with the approach of the critical period, tension continued to mount as prospects for reasonable weather became worse and worse. On the morning of 4 June the predictions received were so bad that Eisenhower reluctantly decided that a postponement of twenty-four hours would be necessary. Worst of all, because of the tides, the latest possible date for the invasion was 7 June.

In the early hours of 5 June the storm reached its height. But the weather forecaster now presented a gleam of hope because a short interval of fair weather was expected which would last until next morning. At this moment Eisenhower was faced with a critical decision. He could take the risk of ordering an airborne and seaborne assault during what was likely to be a temporary and partial break in the bad weather, or he could put the whole operation off for several weeks until tide and moon would again be favourable. He worried that a postponement would be very harmful to the morale of the Allied forces, apart from the likelihood of losing the benefits of tactical surprise. At 4 am on 5 June, with the storm still in progress at Portsmouth, Eisenhower took the final and irrevocable decision: the invasion of Europe would take place on the following day.

On 6 June 1944, the 455 Squadron Operations Record Book notes, 'Today is D-Day – the Allies land in France. It is estimated that over 4000 ships are running a shuttle service across the Channel, and that over 11,000 Allied planes are standing by.' In spite of bad weather, the sea passage across the Channel was successfully accomplished and a degree of surprise achieved for which Eisenhower had hardly dared to hope.

455 Squadron was standing by for work. Six crews spent the day at Manston airfield in Kent, on one hour's notice to fly. They were finally ordered up at 10 pm and made for a patrol area east of the "Overlord" invasion beaches. Bad weather and failing light caused them to split up over the Pas de Calais. Then a crew from 489 Squadron found eight German E-boats.

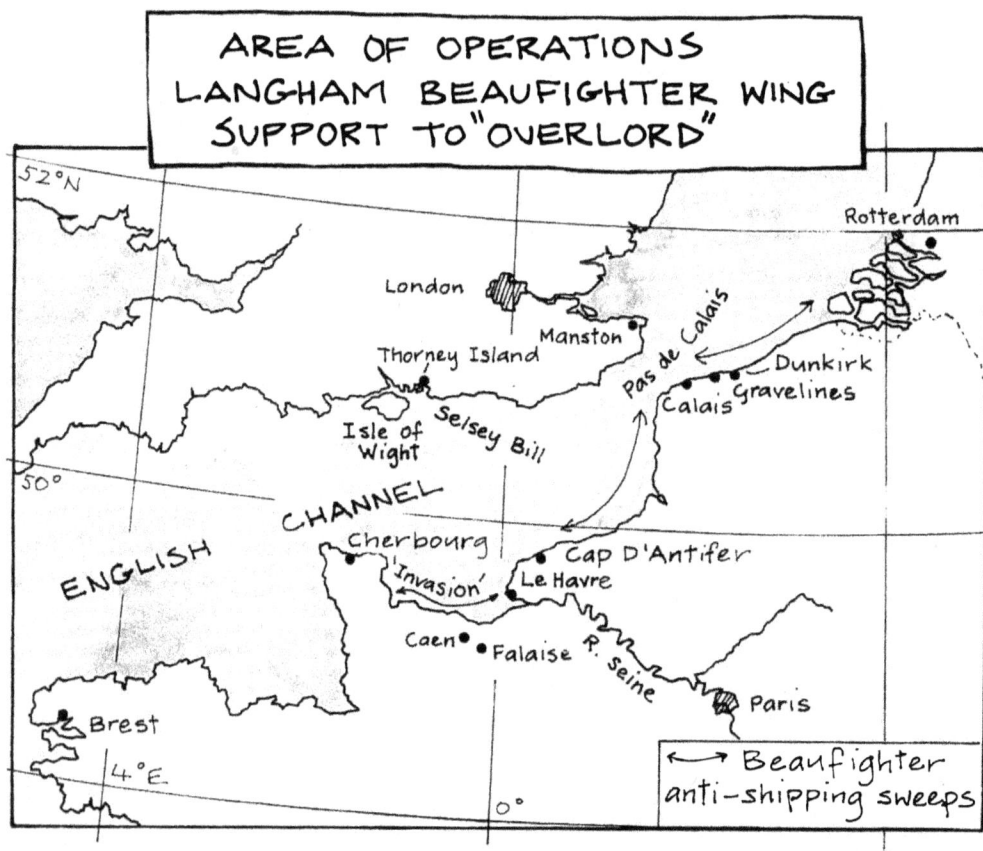

AREA OF OPERATIONS
LANGHAM BEAUFIGHTER WING
SUPPORT TO "OVERLORD"

The E-boat was a fast motor torpedo boat. They were called *Schnell* (fast) boats by the Germans and E (enemy) boats by the Allies. The E-boat's role was offensive. A conservative estimate of the number of E-boats available to the Germans in the Channel on D-Day was 120, most being based at Ijmuiden on the Dutch coast and at Cherbourg on the west Normandy coast. These boats cruised at thirty knots but could sprint at up to forty knots. They displaced 95 tons and were armed with 20 millimetre and 37 millimetre guns. Their practice was to leave port in flotillas and split into groups of two or three in their patrol area, attack convoys with torpedoes then return to base at high speed. They normally operated at night. It was believed by the Allies that the E-boat crews' main fear was from aircraft; not for fear of being sunk but of casualties to personnel. Operations against the E-boats were given the code name 'Conebo'.

Coastal Command found that the 250 pound and 500 pound bombs fused with an air burst pistol were effective against these boats. Although it was unlikely that direct hits would be scored, the blast and splinters from the bomb case which burst just above the water could damage an E-boat up to a hundred metres away.[1]

On this evening of 6 June 1944, the crew from 489 Squadron attacked the E-boats and called all the other crews on the radio, but then lost the targets. There were no more sightings and the attack was recorded as inconclusive. The next day had these crews back at Langham.

Aircraft also stood ready at Langham on this first day of the invasion. A young female driver, Katie Ivatt, recalled those days and her meeting with an Australian airman, Flying Officer Bob McColl.

> I first met Bob out on the airfield where he asked me for a lift back to the Flight Office. I saw him occasionally round the camp after that but our paths did not cross until D-Day. We knew there was something up as all the Beaufighters were painted in weird camouflage paint. Bob had been kicked on the head playing football and, having been to Ely Hospital with concussion, was off flying. He was Duty Officer on D-Day and D-Day+1. I was standing in for my partner on the bus whose husband was waiting on the South Coast for the invasion. So for 48 hours Bob and I woke crews up, took them for breakfast and briefing and out to their aircraft, then waited for them to return and did the same thing in reverse. It was midsummer with long hours of daylight. I remember that towards the end of the 48 hours he said briskly that I could rest my head on his shoulder and doze while we waited for the crews to return but I equally briskly turned down the offer.[2]

The day after the invasion, 7 June, saw better results for the Langham Wing. Just before 3 am six 455 Squadron Beaufighters took off for an anti E-boat patrol in the vicinity of Cherbourg. This time, they found two E-boats and bombed them with 500 pounders - one was damaged. The next afternoon, two Beaufighters left Langham for a patrol along the northern Dutch coast. Flying Officer Fred Williams was flying one aircraft and Flying Officer Neil Smith flew the second. Williams was from Sans Souci, a suburb of Sydney. His navigator, Flight Sergeant Bill Roach was from South Perth. It was their third operational sortie. Off the Frisian Islands Smith received a radio transmission, 'engine trouble, Smithy'. He called back to ask how serious the trouble was but there was no reply. Just then Williams banked to starboard. Smith saw him go and turned to port immediately to follow Williams, who was losing height. Neil Smith recalls,

> he was on my starboard side...slightly below me and right on the 'deck'. There was nothing wrong as far as I could see but within seconds the plane was seen to dip the starboard wing and dive straight into the sea. By the time I turned my plane round to the spot where he had gone in, there was not a thing to see on the water.[3]

The aircraft's reputation for sinking quickly was known by aircrew. One Squadron navigator has commented, 'A Beaufighter could sink in about two minutes. It was a very heavy aircraft without much buoyancy. We were well aware of that.'⁴ Neil Smith circled the area for about ten minutes but saw nothing. Fred Williams and his navigator, Bill Roach were reported missing. Their deaths were finally notified at the end of the War.

The next day the Commanding Officer, Wing Commander Jack Davenport was made a member of the Distinguished Service Order for his skill and leadership during the strike against the German convoy off the Frisian Islands on 14 May 1944. 8 June was also Jack Davenport's birthday. He was 24 years old.

The Beaufighter often proved its ruggedness. The crews also proved theirs. On 11 June, six Australian crews were returning from an anti E-boat patrol along the Dutch and Belgian coasts. They had sighted three small craft that they could not identify so the aircrew turned for base. They reached the coast of England, over Manston in Kent, and set course northwards for Langham. Flight Sergeant John 'Tiger' Payne had his starboard elevator work loose; it was left attached only by the control wires. Suddenly the plane began vibrating, so badly that Payne couldn't read his instruments. Then his plane went into a violent spiral dive but Payne recovered, and with great difficulty climbed to 6000 feet. The plane went into another dive and lost 3000 feet before it was recovered again. Payne's navigator was Flight Sergeant 'Jock' Rennie, a lively Scotsman and a favourite character in the Squadron. He was described as having an accent 'as thick as a bag of haggis.'

Payne climbed again, to 4000 feet. When he reached the Norfolk coast he had Rennie bail out and the navigator landed safely near Great Yarmouth.⁵ Payne was unable to get out - he had to keep considerable pressure on the aircraft controls. He is credited with recovering from seven spiral dives during that trip home. As Payne reached Langham he decided to have a go at a landing. He made it at over 180 knots, approximately twice the normal speed. The Squadron Record Book records his 'skilful airmanship'. Corporal Reg Andrew, a Squadron armourer, recalled Payne's landing. Andrew says, 'When he got back, he sat on the ground, hunched against the landing wheel, shaking all over.'⁶ The Commanding Officer, Jack Davenport, recommended Payne for the Immediate Award of Distinguished Flying Medal. The citation says,

> Flight Sergeant Payne, displaying outstanding courage, determination and flying ability, brought his aircraft back to base and successfully landed it without causing further damage...The keenness of this N.C.O is outstanding, and his resolute determination and ability are an example to all.⁷

The award was made to Payne shortly afterwards, for his effort that day and for his previous work.

On 12 June 1944 the Battle for the build-up in Normandy continued. Field Marshal Irwin Rommel, in command of German Army Group B, summed up the situation in a signal to Field Marshal Keitel at German High Command, 'The enemy is strengthening himself visibly on land under cover of very strong aircraft formations. Our own air force and navy are not in a position to offer him appreciable opposition - especially by day.' At a

'Attached only by the control wires': The damaged starboard elevator on the Beaufighter flown by Flight Sergeant John 'Tiger' Payne when he returned from an anti E-boat patrol on 11 June 1944. Payne had his navigator bail out, then struggled to bring the almost unmanageable aircraft back to base at Langham. (Photo RAAF Historical Section)

conference in Berchtesgaden, Grand Admiral Doenitz, Field Marshal Keitel and Colonel Jodl reached the conclusion that 'if the enemy succeeds in fighting his way out of the present bridge-head and gains freedom of action for mobile warfare, then all France is lost'. There was one strategy that might induce the Allies to attempt a second landing in northern France, one that might interrupt the build-up at Normandy. That strategy was a bombardment of London by flying bombs, beginning that night. Although only one bomb reached its target as planned on the 12th, the strategy remained in place.

This was a vital time for the Allies, and Coastal Command's aircrew were on a demanding schedule. A 455 Squadron pilot wrote at the time,

> ...we have to remain on the camp all the time. I came off standby at 11.30 am having been on since 4.55 am this morning. I managed to get approximately three hours of sleep in this afternoon. That's all we do nowadays. Go on standby, come off, eat, sleep and do another standby or trip. I won't be surprised if I'm hauled out about 11 pm or thereabouts tonight. We seem as though we are going to keep a grip on the beachhead in France.[8]

'Recovering from seven spiral dives': Flight Sergeant John Payne from Lismore, New South Wales, who flew with 455 Squadron from November 1943 until March 1945. He was awarded the Distinguished Flying Medal for his action on 11 June 1944 and for his part in many anti-shipping operations with 455 Squadron.

During the brief spare time that was available in those busy days, Flying Officer Keith Carmody had some 455 Squadron members help him lay a concrete cricket pitch on the sports field at Langham. Carmody was one of a number of talented and prominent cricket players in the Squadron. Less than a month earlier, four Squadron members had played in an unofficial Test Match between an Australian Services XI and 'The Rest'. There had been plenty of talent at the match. Included in 'The Rest' team were Len Hutton, Wally Hammond, Bill Edrich and Dennis Compton. The Australian XI had, from 455 Squadron, Flying Officer Carmody (Captain), Flight Sergeant Bill Roach (killed on 8 June 1944), Pilot Officer Bob Cristofani and Flight Lieutenant Roper. Guests at the match were the Australian Prime Minister John Curtin and Australia's Commander-in-Chief, General Thomas Blamey.

Keith Carmody brought enterprising captaincy to his team. He had invented what became known as the 'Carmody field'. Other captains had used several men in slips but not in the umbrella formation Carmody used to pack his slips area with catchers. Flying Officer Carmody was on an anti E-boat patrol with 455 Squadron on 13 June 1944, days after laying the Squadron cricket pitch.

Also on patrol that day was Warrant Officer John Ayliffe from Peterborough, South Australia. John had been a clerk with the South Australian railways before the War, and began his aircrew training at Victor Harbour in December 1941. After over two years training in Australia, Canada and Britain, he finally arrived at 455 Squadron just before it left Leuchars, in March 1944.

Langham and North Coates Wings both sent two strike squadrons on this anti E-boat patrol. It was a time when the campaign against the E-boat was reaching its peak. 455 put up twelve aircraft, which was close to maximum effort for a squadron which normally had only twenty Beaufighters on strength – there were always a number unavailable because of essential maintenance. Jack Davenport led the Australians again. They left before first light, at 3.45 am, flying south to Gravelines on the French coast near Calais, then north-east up the Belgian coast towards the Hook of Holland. Off the island of Schouwen on the Dutch coast, just short of the Hook, and now at first light, they found an armed trawler and two smaller vessels. Jack Davenport took Warrant Officer Noel Turner with him to attack. Noel was from Hurstville, New South Wales, and had joined the RAAF early in 1941, arriving at 455 Squadron in January 1944. On their run-in they came under heavy fire from the escort ships. Noel Turner recalled this strike,

> (The flak was) very impressive. I knew that no aircraft could fly through it and (I) flew unhappily for a while along the fringe. Then at a possible gap I turned the aircraft on its side and flew in. Got the starboard wing clipped and aircraft vibration was severe. Not game to take evasive action for fear of a wing falling off, we popped out of the convoy area like a cork from a bottle at sea level. In some strange manner F/S Costello who had been right behind me prior to the attack was able to come on air and ask if I wanted escorting to base. 'No, we'll be OK thanks' . Davo [Commanding Officer] had had enough of this. 'Leader to Red Two, rejoin Squadron and continue patrol.' 'Red Two to Leader. Aircraft damaged. Returning to base.'9

The Squadron Operations Record Book records that Noel Turner scored 'a direct hit on one vessel'. However, Noel says that although he has a clear recollection of taking aim and dropping his bombs, he didn't see the bomb blast because he 'was busy making a turn and getting out of there before the explosion'.

While Noel Turner made his way back to Langham, 455 Squadron continued up the Dutch coast. Off the Hook of Holland they found a convoy of seven armed trawlers and they flew back out to sea to rendezvous with the formation from North Coates. The Australians jettisoned their bombs and both groups came back for an attack with cannon. On the opposite side, flak was heavy again from the ships and shore batteries. Milson led a section of six Beaufighters in the attack; they kept firing 'down to deck level'. Five of the seven trawlers were left severely damaged and on fire and the other two were claimed as damaged. However, five aircraft were hit by flak. One of them was flown by Flying Officer Carmody, who radioed that he had engine damage and would have to ditch. His aircraft was last seen with smoke coming from an engine and heading towards the coast some five kilometres away. Carmody ditched and he and his navigator spent twenty-one hours in their aircraft dinghy before they were picked up by a German R-boat.[10]

Noel Turner had managed to get his damaged Beaufighter back to Langham. He says,

> Never had J-Johnny been handled so tenderly as on that return trip. When we parked it, Wiggy [Squadron Leader Wiggins] came strolling over to survey the damage. 'Why didn't you get your navigator to shoot off the damaged section' he said with a grin, 'you could have continued your patrol then.'[11]

Noel was to become one of the most experienced pilots in the Squadron. For his action in this attack and his later work he would be awarded the Distinguished Flying Cross.

Warrant Officer John Ayliffe's Beaufighter was another of those damaged. He says, 'The flak seemed to me to be coming between the fuselage and the engine nacelles. On my return to base I experienced difficulty in lining the aircraft to the runway...After parking the aircraft I counted about fifty holes in the wing and fuselage.'[12]

Keith Carmody and his navigator, Flying Officer Gilbert Docking, were posted as missing. This had been their eighth operational sortie with the Squadron. They were later reported as prisoners of war and were held at Stalag Luft III in Germany. Stalag Luft III was the Prisoner of War camp where, less than three months earlier, Squadron Leader Jim Catanach had escaped, was recaptured, and shot along with forty-nine other men. Carmody and Docking were freed late in the War when the Russian forces reached the outskirts of Berlin. Keith Carmody finally arrived back in England just as the Australian Services Cricket XI was formed for the Victory Tests.

Flying Officer David Whishaw was one of the youngest pilots in 455 Squadron, turning twenty-one years old in April 1944. He had joined 455 Squadron at Leuchars in April, earning his Wings in Canada through the Empire Air Training Scheme. By mid-June Whishaw had already taken part in several attacks on shipping and in many anti E-boat patrols. He recalled his time at Langham,

> Station transport was always available for operations or other duty, but we had to find our way from our huts to the Mess and elsewhere, which was some distance, on foot. Eventually we were all issued with bicycles which soon tended to degenerate into playthings. I shared a hut with Bob McColl, just up the road from one occupied by Westralians, Colin Cock and Wally Kimpton. Bob and I rode down one day to find Col and Wally, stony faced, trying to fling their bikes over their hut. This craze soon caught on and on the rare occasions that someone got a bike over the top and down the other side there would be shouts of approval. Sometimes a bike was left on top. Of course all this resulted in bent pedals and handle bars and in extreme cases, buckled wheels and for a while there would be

'double-dinking' from place to place on the remaining serviceable bikes.

The Mess was usually overcrowded, kitchen facilities basic and the food, though abundant, was invariably terrible. We suspected darkly that horse meat was served at least once a week. We made best use of parcels from home and from the Red Cross. Most of us smoked and the bar was well stocked with cigarettes, the only acceptable English brand being Players, but we also got a monthly issue of a 200 pack of American Lucky Strike, Chesterfield, the extra long Pall Mall or the extra strong and unfavoured Camels, said to come from the nether end of those animals. For non-smokers, this issue could be readily traded, sometimes at nearby farms in exchange for delectable items of produce. Spirits and ales were available, but for economy reasons and a long Australian beer drinking history, 'mild and bitter' was the common beverage, usually described as 'piss'. Since flying was on in some form most days, heavy drinking was rare except when there was a party for some reason or another.

During one of these, the Mess was suddenly invaded by three medium-sized pigs. Anyone familiar with pigs knows they are heavy, slippery and utter deafening squeals when apprehended. The idea was to try and jump aboard a pig rodeo style, but this met with little success and extreme protest from the pigs. How they were caught, stolen and transported, remained an extraordinary mystery. I think we finally believed their owners must have been bribed heavily for removal from their sty, or perhaps blackmailed by threats of withdrawing supply of swill, which would have been a bonanza for local pig farmers.

Life could get a bit tedious between flying, particularly when the weather closed in. Snooker and table tennis tournaments were often organized and Ted Watson, the 'elder' of our group who taught some of us to play contract bridge. Amongst the regulars was 'Chick' Smith, ex-commercial traveller, a cheerful individual and always an entertainer. Chick, like many quite hopeless bridge players, still loved the game. He could never resist letting everyone know what was in his hand. When dealt poor cards he would squint at them - 'Not worth two squirts of tiger's piss'.

Usually though, it was a constant round of air tests, air exercises and operations. At this time huge armadas of USAAF Fortresses and Liberators were constantly on daylight bombing missions and sometimes set course within sight of Langham. It might take up to an hour for them to circle, gain their height of over 20,000 feet and form up. It was an awesome sight to watch them finally set course east. One morning we heard at breakfast that a Fortress had made a forced landing at Langham at dusk the previous evening. Some of us got a jeep and drove out to a far perimeter of the drome where the aircraft stood. Miraculously the tyres and undercarriage were undamaged, but two propellors were feathered, a gun turret and part of the rudder was missing and the wings and fuselage were extensively holed and torn. Watson's navigator, Vic Smith climbed in and almost immediately climbed out again white-faced, saying 'There's a stiffy in there and there's blood everywhere.'

Thirty years later I went back to Langham. Over the whole aerodrome came a low humming sound which emanated from modern pens housing thousands of turkeys served by the runways and perimeter tracks. Everywhere a stench arose. All the Nissen huts and other buildings were gone except the control tower and half our old Mess standing roofless and the rest a heap of rubble. On a still summer's day it stood isolated, deserted, silent. How could anyone believe it was once a powerhouse of youthful life and endeavour. I could not have been there at night for thought of imagining the ghosts of companions long gone - Barbour, Kempson, Cock, many others; the shambles of that place offensive to their memory.[13]

North Coates and Langham had continued planning the new combined strike wing tactics. Air Vice Marshal Ellwood, the Senior Air Staff Officer at Headquarters Coastal Command said of these tactics, 'We learnt the unwisdom of attacking in small numbers. The bigger the force the more complicated is the problem of the A.A gunner who, amidst a hail of bullets, is hard put to it to pick his target.'[14] After the joint wing exercise on 31 May, it was time for the tactics to be used. The Coastal Command Review noted,

> On 15 June, another foundation stone had been laid in the ever-changing structure of anti-shipping tactics. The enemy, who had gained a healthy respect for the attacks of our Beaufighters, had continued to increase the escorts to his precious merchant ships. Our counter-measure was to send a combined wing strike to attack such convoys.[15]

Allied Intelligence found that two important and newly completed vessels were in transit from Rotterdam to Poland. One was a merchant ship of 7,900 tons known called the *Amerskerk*. It had been laid down in Amsterdam, to Dutch order, and after launch in November 1942 had been towed to Rotterdam for fitting out. On the way to Rotterdam it was attacked by aircraft of Coastal Command. It was repaired and fitted out, and in the weeks leading up to 15 June had been seen in several ports in Rotterdam, suggesting that it was ready for service. The other vessel was the 3,500 ton *Gustav Nactigall*, also on its maiden voyage. It had been built at Hoboken, Belgium, to Polish order as a merchant ship. After the occupation of Belgium the *Nactigall* underwent a lengthy transformation to equip it as an E-boat depot ship and it moved to Antwerp in May 1944, then to Rotterdam where it had been last seen on 12 June. These ships were to move in stages up the coast, and were expected to be heading for the port of Gydnia near Gdansk on the Baltic Sea.

The German convoy had sailed from Den Helder at 9.55 pm on 14 June and was heading for its next anchorage at Borkum. It was led by three minesweepers of the German 7th Flotilla in an arrowhead formation. Behind them was the *Nactigall*, heavily armed with two 105 millimetre guns, two 37 millimetre guns, and ten 20 millimetre cannons. In the wake of *Nactigall* was the *Amerskerk*, armed with eight 20 millimetre cannons. On the flanks of the formation were four more minesweepers of the 7th Flotilla. Each mine-sweeper carried two 105 millimetre guns and six cannons, and most were also armed with machine guns. By dawn of the 15th, the convoy had made good time and was nearing the Dutch Frisian island of Schiermonnikoog. There was a westerly wind of about 25 kmh, a moderate sea-swell and good visibility.

16 Group planned a strike for the morning of 15 June. For the first time, two strike wings were briefed to attack the two ships and their escorts. The

force was to be the largest ever sent into action by Coastal Command. The North Coates Wing assembled ten torpedo Beaufighters and four anti-flak Beaufighters from 254 Squadron RAF and five Beaufighters from 236 Squadron RAF armed with rockets and cannon. The Wing at Langham provided twelve cannon-firing Beaufighters from 455 Squadron and eleven cannon-firing Beaufighters from the New Zealand 489 Squadron. 455 Squadron was to be led by Squadron Leader Colin Milson. One of Milson's pilots was Flying Officer Ted Watson from Drummoyne in Sydney. After leaving high school Watson became a teacher and joined the RAAF some nine months later. He began flying training in December 1941 and went solo after thirteen and a half hours flying. Watson completed aircrew training in Canada with a number of other aircrew now with 455 – Bob McColl, Dave Whishaw and Frank Proctor – and he arrived in England in May 1943. Ted Watson was posted to 455 Squadron at Leuchars on 27 August 1943.

This force of forty-two Beaufighters was escorted by ten Mustangs of 316 (Polish) Squadron stationed at nearby Coltishall. The whole force was to be led by Wing Commander Tony Gadd, the Wing Commander Flying at North Coates. North Coates and Langham were only fifteen minutes flying time apart. Forming up such a large strike force required precise timing and careful flying, especially since the squadrons had to take off in the early morning darkness to make the interception with the convoy before it reached shelter in a defended anchorage.

First the two strike wings had to link up. The North Coates Wing took off at 4.10 am and set course south-east for Langham. The wings merged over Langham and headed for the coastal town of Cromer. Near there they were joined by the Polish Mustangs. Flying Officer David Whishaw was flying Beaufighter UB-P, his usual aircraft. He recalls,

> Some of the planes had seen a lot of service. P was a dreadful old bomb which would not trim to fly straight and level, so that one had to frequently correct course and altitude and most irritating of all, revolutions of the two engines constantly went out of synchronisation. To cap all of this, P was the slowest in the Squadron. I soon called her 'P-piss poor' and this was conceded as being an accurate description by any of the other pilots who had the misfortune to fly her. I was stuck with P from January till 15 June. I clearly remember that early and dark morning as the 455 aircraft lined up the runway in tandem for take-off...Having formed up and set course over the Norfolk countryside, the port motor of P started backfiring and belching smoke. There was nothing for it but to drop out of the formation and head back for Langham. This was utterly frustrating, but it was the last the Squadron saw of that particular P, which was then replaced by one of better quality.[16]

The force set course for Ameland in the Dutch Frisian Islands, 250 kilometres away. Out in front were the North Coates rocket-equipped Beaufighters, with the New Zealanders on the port flank and the Australians slightly further away to starboard. The North Coates torpedo Beaufighters were behind. The Mustangs in escort flew above and behind the Beaufighters.

After leaving the Norfolk coast, Gadd brought the force down to just above sea level, below the German radar, and the wake formed in the sea by the propellors of the aircraft stretched back towards the English coast. It was a large formation and keeping station was important. Each section was itself made up of groups of three Beaufighters, called Vics, a leader and one on each side of him, slightly behind. Pilots drove their aircraft close to their leader's wing with throttle and joystick, wing-tips often less than three metres apart.

The strike wings flew across the North Sea and into thick fog and cloud. As they lost touch the squadrons took action according to their plan. The leading squadron maintained course but began a gentle climb. On either side, the Australians and New Zealanders opened out, each by ten degrees and they also began to climb. They would fly like this for five minutes, then resume their original course. When they broke free from the cloud they were to look for each other.

After about half an hour the fog and cloud began to clear and as they flew into the sunshine they saw parts of the Dutch and Frisian islands stretched out ahead of them. Visibility was good, a bright day with 6/10ths strato cumulus cloud at 1500 feet and a slight sea haze. At 5.56 am the convoy was sighted off the island of Schiermonnikoog. It was spread over a large area; each of the major ships was closely flanked by an escort, their wakes prominent as they attempted to maintain speed and position. The Torbeaus of 236 Squadron had moved ahead and to the left of the main formation. Gadd knew he would have to put the anti-flak sections in position to attack first and synchronise their attack with the Torbeaus. He radioed to all aircraft that he was turning to starboard. Gadd's navigator was Flight Lieutenant Duncan Marrow who recalled,

> Tony [Gadd] manoeuvred for position but we were just a little too near the convoy. It takes a hell of a lot of sky to bring 40 odd aircraft through ninety degrees. The formation on the right, which was the inner part of the 'wheel' could not throttle back sufficiently to keep in formation. So the leader [of the right section] took whole force beneath our force and Tony slipped the middle force to take his place on the right side. By that strategy all the Beaus were more or less in line abreast, with the exception of the torpedo boys.
> Paddy Burns [commanding 254 Squadron RAF, carrying torpedoes] had gone straight through the cloud to sea level or thereabouts. The convoy spotted his force and started firing at them. He got quite close to the ships, almost within dropping range of the torpedoes, then wham - through the clouds came about three dozen aircraft with guns a-blazing.[17]

The 455 anti-flak section swept low over the convoy strafing the escorts. Cannon strikes kicked up the water all about the ships and aircrew saw the ships' crews firing back. Squadron Leader Colin Milson said,

> As we came down, firing continuously, we split up into small strafing groups and each group picked out a minesweeper or two to beat up. One of our pilots found he couldn't get his Beau into position to attack a minesweeper so he took on the big merchant ship which appeared to be filling the whole of his skyline...I went down on to a minesweeper with two aircraft beside me and a couple following

'We split up into small strafing groups': Squadron Leader Lloyd Wiggins DSO (left) strafes a minesweeper with 20 millimetre cannon during the combined wing strike off the Frisian Islands on 15 June 1944. Squadron Leader Colin Milson DFC followed and took this photograph of the strike. A balloon tethered by steel wires can be seen above the minesweeper.

behind who were lobbing their shells uncomfortably close. They were passing over and under our wings as we closed our range to the minesweeper... I suppose we gave that ship at least an eight second burst before we swung away at deck level...When we swung away from her we climbed steeply and dived on other ships in the same way.[18]

The Torbeaus followed, scoring hits on each of the two major vessels. The escort ships also suffered. One M-Class minesweeper attacked by Milson and Flying Officer Colin Cock was so badly damaged by cannon fire that it blew up and sank. Milson says, 'While the action was going on the Poles who were escorting us in their Mustangs became very excited and we could hear their comments over the R.T. as they cheered us on from their grandstand seat'.[19]

The Beaufighters gathered into loose squadron formations and headed for base. Flying Officer Neil Smith had the collar of his Mae West jacket torn by shell splinters. His plane had its starboard section badly holed, his hydraulics were shot up, and part of the cockpit was holed but he made a successful belly landing back at Langham. Colin Milson said later, 'of all the things I remember best about it was the wide grin on the face of an Australian pilot Doc Watson - it was the first time he had had a chance to get loose with his cannons and he took full advantage of the occasion. On his way home his aircraft felt a bit wobbly. He didn't realise it at the time but an enemy cannon shell had gone through his engine and a piece of shrapnel through his tyre. But even that didn't spoil his joy'.[20]

After the strike was over it was clear that the new tide in strategy had been prosperous for the attackers. It had been the largest and most successful strike ever laid on by Coastal Command, and both the North Coates and Langham Wings were congratulated by the Air Officer Commanding Coastal Command. Not a single aircraft was lost. The Coastal Command Review concluded, 'The chief reason for this very fine victory was the presence of a really adequate anti-flak force which thoroughly discouraged the enemy gunners.'[21] A reconnaissance flight to the area a short time later reported five balloons flying above the water, still attached to ships lying on the bottom of the North Sea. The *Amerskerk* and the *Nactigall* were both lying on their sides, awash. Nine Germans were killed in this attack, one was missing and another seventy were wounded. After this operation, Squadron Leader Colin Milson was recommended for the Immediate Award of a Bar to the Distinguished Flying Cross. The recommendation read,

> His section was first in to attack, and again, with his consistent determination, oblivious to the heavy and light flak, he pressed home his attack on two of the escort vessels, completely silencing all opposition. This Flight Commander is an outstanding personality. He is a brilliant leader, possesses high administrative ability, and his courage, eagerness to meet the enemy, and determination, are of a standard all pilots should emulate, and are demonstrated by the fact that on every strike in which he has participated, he has attacked not one but two or three of the enemy ships, and he is indeed an inspiration to all.[22]

That evening the intense flying-bomb offensive against London began. In the first twenty-four hours, 244 flying bombs were launched, and 73 reached London, killing more than fifty civilians. To prevent civilian panic, the British Government ordered that the number of newspaper obituary notices for people killed by enemy actions should be limited to three from the same postal district on any one day.

The responsibility for keeping the E-boats in check rested with 16 Group Coastal Command and naval forces. On 16 June, Langham strike wing crews were on standby as usual. One of the 455 pilots was Flying Officer Ted Watson. At mid-morning six crews were scrambled and Watson was first away. He flew to the patrol area near Ostend and there each pilot patrolled separately. Watson made the first sighting at 11.37 pm. His diary records,

16 June 1944:

Saw one wake but lost it manoeuvring for position. Continued patrol and saw four more - dropped 4 bombs followed by accurate A.A. Two probably hit as Jack Payne who followed saw only two wakes and Bill Jones after him only one wake. No flak at all when we went in.[23]

Debriefing at RAF Station Langham after a strike in June 1944. Aircrew are, standing from left to right: P/O Alec Jones RAF, S/Ldr Lloyd Wiggins DSO, F/Sgt Harold Morris RAF, F/Sgt Ted Holmes RAF (with cigarette), F/O Wally Kimpton, F/O Frank Proctor, F/O Fred Dodd RAF, F/O Laurie Jones RAF, F/O Bill Barbour, F/O Neil Smith, F/O Jack Belfield RAF (nearest camera), F/O Forbes Macintyre.
Seated, left to right: Intelligence Officer, Commanding Officer W/Cdr Jack Davenport DSO DFC, F/O David Whishaw (partly obscured),(not identified), F/Lt Ian Masson DFC RAF.

The flying bomb attacks on London were continuing, although in the last fortnight of June 1944 the Germans were able to launch only 2000 flying bombs, of which barely a thousand reached London. Allied bombers attacked the launching sites and severely restricted the scale of the attack. On 18 June Eisenhower was assured by Churchill that there must be no question of changing Anglo-American strategy to achieve an early conquest of the flying bomb bases. The German strategy of interfering with the Normandy build-up failed. Montgomery had twenty divisions ashore, opposing a German force of fourteen divisions. With the Normandy beachhead secure, the Allies had to break-out. The place for that was the town of Caen, about twelve kilometres inland from the invasion beaches. The date for the break-out was fixed for 26 June.

Since the large combined wing strike on 15 June and the anti-E-boat patrols on 16 June, 455 Squadron had a quieter period for nearly two weeks. It appeared to the Squadron that Coastal Command's vigorous attacks of early June had kept the enemy ships out of the waters just north of the English

'Imperturbable Bert': Navigator F/Sgt Bert Iggulden RAF shows his pilot, F/O Steve Sykes, the fragment of flak that lodged in his flying boot during an attack on German minesweepers near the Hook of Holland on 28 June 1944.

Channel. Air Chief Marshal Sholto Douglas began to turn his anti-shipping strike wings against German and coastal merchant shipping – slower and less manoeuvrable targets that offered a greater result from air attack.

The temporary lull ended on 28 June 1944. That day there were five separate patrols by the Australians, involving eleven aircraft. It was a very early start; the first sorties were away before dawn, at 2.15 am. Squadron Leader Colin Milson led six 455 Squadron Beaufighters and six from the New Zealand squadron in a patrol from Dunkirk north-east to the Hook of Holland. The two sections lost contact on their way to Dunkirk and continued their patrols separately. Near the Hook the Australians found seven vessels they believed to be German minesweepers. They attacked immediately in a shallow dive from 1500 feet. The Germans replied with intense flak and rockets. Flying Officer Steve Sykes from Currawang, New

South Wales, had his plane damaged and his navigator, Flight Sergeant Bert Iggulden was hit in the foot by a fragment from the nosecap of a 20 millimetre shell which lodged in his boot.

New Australian aircrews continued to arrive at Langham. Before flying on operations, the new crews were sent on a number of local navigation exercises and, with other experienced operational pilots, they practised low-level formation flying. On 26 April 1944, Flying Officer Ted Collaery and his navigator Flight Sergeant Vic Pearson had arrived at Langham from Number 9 Operational Training Unit at Crosby-on-Eden, in Wales, to join 455 Squadron.[24]

Ted Collaery was from Wollongong, New South Wales. After finishing school he had studied wool classing by correspondence. He was an accomplished horseman, studious, with a love of literature, and he often rode long distances from Wollongong to a small hut the family owned in Kangaroo Valley near Nowra, where he kept a supply of books. Ted had joined the Air Force late in 1940, when he was twenty-six years old. He began his flying on Tiger Moths at Narrandera in February 1941 and continued training in Alberta, Canada. He spent nearly two years flying at an air gunnery school at Evarton in Scotland before he began his conversion to Beaufighters. In March 1944, at the Operational Training Unit, he had teamed up with his navigator,

Breakfast at RAF Station Langham for crews from 455 Squadron after an anti E-boat patrol off the coast of Holland on 28 June 1944. They departed base at 3.30 am, attacked seven German minesweepers near the Hook of Holland then returned to Langham at 6 am. Seated around the table are, from the left clockwise: F/O Colin Cock (pilot), F/O Clive Thomson (pilot), S/Ldr Colin Milson DFC and Bar (pilot), F/Sgt Bob Taylor (navigator), F/Sgt Bert Iggulden RAF (navigator), F/Sgt Ivor Gordon (navigator), F/Sgt Bob Lyneham (navigator), F/Sgt Jack Costello (pilot), F/O Lee Turner (navigator), F/O Steve Sykes (pilot), F/O Bob McColl (pilot) and F/O Matt Southgate RAF (navigator). (Photo courtesy S. Milson)

Vic Pearson. By the time Ted Collaery left the operational training unit, he had logged over eight hundred hours flying with over fifty-five hours on Beaufighters.

Vic Pearson was from Goomeri, near Maryborough, Queensland. He had left school aged fourteen and his father had been happy to have him to help on their farm. Vic went to Gatton Agricultural College and then went jackerooing in central Queensland. When the war started Vic joined the Air Force and although he wanted to be a pilot, his school results weren't considered good enough and he was mustered as a wireless-air gunner. He saw it as a stroke of luck when his whole course was sent to do navigator training. With many other Empire Air Training Scheme aircrew, Vic was sent to Britain via San Francisco, across the United States by train, and then across the Atlantic on the *Queen Mary,* with 14,000 other men. His progress in England was also typical. From the Personnel Reception Depot, Vic Pearson did the reconnaissance course at Squires Gate near Blackpool, then to his operational training unit at Crosby-on-Eden. There he met Ted Collaery. In 1944 Vic Pearson was 31 years old.

Their early days at Langham had been very busy. Shortly after they had arrived at Langham Ted Collaery wrote to his wife, 'The CO said he was going to have us operational in the shortest possible time and he seems to be living up to his word for we fly day and night. It was 1 am this morning when I got to bed and at 9 am was at briefing for a navigation trip from which we landed at 3 pm.'[25] Collaery's and Pearson's first full operational sortie was an anti-E-boat patrol on 20 May, a night operation lasting 2 hours 45 minutes.

Late in June 1944, Flying Officer Jack Cox and his navigator, Flight Sergeant Allan Ibbotson from Kalgoorlie, Western Australia, were also new to the Squadron. They had arrived from Number 36 Operational Training Unit at East Fortune on 16 June 1944. Jack was from Dungog, New South Wales. He had been share farming in 1941 before he joined the RAAF with the Empire Air Training Scheme and had trained in Canada and England. He flew Hudson and Wellington bombers before volunteering to fly Beaufighters in 1944. When they arrived at Langham they were allotted to a Flight, to their accommodation, and issued a .38 Smith and Wesson pistol, a push bike and a parachute. Jack Cox recalled,

> Our first impression was one of peace and quiet. No flying was taking place but when we visited the drome we found the reason - every serviceable aircraft was on a dispersal point, fully fuelled and armed, with a crew aboard and ground staff standing by. 'Operations' were continually receiving reports of Channel activity and a 'sighting' on enemy shipping was signalled by a Very pistol from the control tower. Engines immediately sprang to life and as they taxied to take off, target and position were indicated on a large black board at take-off point. On 22nd June things were quiet enough for four aircraft to be made available for W/C Davenport to take us on the customary introduction to low-flying formation.[26]

Jack says that they flew at low level with the aid of a radio altimeter. It lit a red warning light according to the altitude set on the control. He continues,

The Commanding Officer of 455 Squadron, W/Cdr Jack Davenport briefs his crews before anti E-boat patrols from Langham in June 1944.
Standing in foreground: W/Cdr Jack Davenport DSO, DFC. Standing on the left: F/Sgt Albert Vigor (partly obscured), F/O Ted Collaery.
Seated, front row, left to right: F/O Steve Sykes (partly obscured), F/O Ted Watson, F/O Vic Smith RAF, F/O Lee Turner, F/O Forbes Macintyre (map on knees), F/O Jack Cox, F/Sgt John Ayliffe, F/O Neil Smith, F/Sgt Allan Ibbotson.
Seated, back row, left to right: P/O Vic Pearson, F/O Fred Dodd RAF, F/O Bill Barbour, F/Sgt John Payne DFM, F/O Leo Kempson, F/O Wally Kimpton, F/Sgt George Kerr RAF.
All ranks are as in June 1944.

'Some pilots set a height and flew with the light flickering. [This day] we flew in formation with three other aircraft at what I thought was fairly low level. When we got back to base Wing Commander Davenport said to me, 'If you ever fly that high again, I'll shoot you down myself'.

Jack Cox's first attack against shipping came soon, on 29 June 1944. Collaery and Pearson also flew that day, their eighth operational sortie. Early that morning, six 455 Squadron Beaufighters teamed with five Beaufighters from 489 Squadron for another anti E-boat patrol from Dunkirk to the Hook of Holland. The eleven aircraft took off from the Langham field at 3.40 am to catch E-boats returning to port after night attacks. Flight Lieutenant John Pilcher led this patrol, armed with two 250 pound bombs. Visibility was poor and flying at one thousand feet they followed the French coast eastward towards Holland. At dawn they found an enemy convoy of two minesweepers, two trawlers and nine other small unidentified ships making for harbour at the Hook of Holland. Flying Officer Jack Cox recalls their reaction, 'Our bombs were useless against such a target and were hastily jettisoned as we turned and climbed to starboard to position ourselves for the attack with cannon.'[27]

Pilcher had positioned Jack Cox on the outer side of the formation, furthest from the shore defences, where life was supposed to be a little easier for new crews. When the formation of aircraft turned to make its attack, the German ships had reached the cover of heavy shore flak batteries. Jack Cox prepared to attack. He says,

> I wheeled around with the formation and that put me closest to the shore defences. As the formation straightened up we got the order to attack and my plane was hit straight away. The leading edge of both wings were smashed by the high explosive flak. I lost all my instruments except the compass and air speed indicator.[28]

Jack Cox had committed his plane to the attack dive, so he pressed on. He was hit again by flak which exploded near the cockpit and lodged a piece of shrapnel in his arm.

> I had four ships to my front and I fired my cannons on our way in. By the time we were on to the third and fourth ship we'd come down so low that we were looking up at their bows. I headed for the open sea but we started to take hits on the wings from these boats also. I was too low to take evasive action and the German gunners were directly behind me so they didn't have to aim off. All I was thinking about was getting as far from the convoy as I could before ditching. I could see the trailing edges of both wings being hit. Something made me think 'weave, weave'. I gained a little height and the tracer bullets followed me. I flew lower again and found that it took the gunners several seconds to adjust their aim. I had to repeat this manoeuvre several times before we were out of range. I felt quite elated with the effectiveness of this and gained height.[29]

In the navigator's seat, mid-plane, Allan Ibbotson was unaware that his pilot had been injured but was having troubles of his own. He says, 'It was obvious that we'd suffered a lot of damage and Jack asked me if I wanted to bail out. My hatch was damaged which would have made it impossible so I told Jack we should press on.'[30] Jack Cox did that. He says,

> Although I had no indicators for fuel, oil or engine revs, we decided to try to make it back to Langham rather than use one of the emergency strips. I concentrated on keeping the plane straight and level, never giving a thought to what we'd do if we reached Langham. On seeing the drome I flew out over the 'Wash', turned, lined up the drome and called flying control saying 'I'm coming straight in'. I knew I had to crash land on my first attempt - no flaps and not prepared to lower the undercarriage - (I don't think flapping wheels would have helped very much.) I overshot slightly, more than I intended, and the drome was sloping away - I had to lower the nose to keep close and level to the ground, building up the airspeed and carrying me further down the drome. This carried me beyond the tarred runways and hard ground and I eventually touched down in a patch of cabbages, (cultivated by Italian POWs). The landing was comparatively smooth, I undid my harness, jettisoned the canopy and was standing on the wing when the ambulance arrived.[31]

Jack Cox became used to belly landings. He says, 'In my flying career I crash landed four aircraft - two very overladen Hudsons and two Beaufighters. The Beaufighter wasn't hard to crash land. It was just a matter of flying it flat and

'Something made me think "weave, weave"': Flying Officer Jack Cox was wounded by shrapnel in a strike against German shipping on 29 June 1944. Jack Cox was awarded the Distinguished Flying Cross for pressing his attack despite his wounds and the damage to his aircraft.

level onto the ground.'[32] Not surprisingly, the Squadron didn't use Beaufighter UB-J again. For his efforts in pressing the attack and bringing his plane back, and for later work with his Squadron, Jack Cox was later awarded the Distinguished Flying Cross. From this episode he earned the name 'Flak Jack' from his friends in the Squadron.

Another crew suffered badly in this attack. Flying Officer Ted Collaery's Beaufighter was hit by flak, damaging the starboard engine. Collaery reported the hit, feathered the engine and set course for base. He was now free from the flak but not from danger. The Squadron Operations Record Book says,'...though he maintained height on one engine for a short while, he was soon compelled to ditch'. Navigator Pearson recalls, 'I had managed to get an SOS out before we ditched and gave our position.'[33] Pearson got out of the aircraft and launched the dinghy. He says, 'I went forward to the pilot's hatch to help him get out...he was struggling to get free. I don't know why I couldn't get him out. We may have taken a hit in the nose that damaged the cockpit where his feet were.'[34]

The plane was sinking and Collaery was fighting to be freed. The Commanding Officer Jack Davenport later wrote that, 'F/O Pearson

'I'm coming straight in': Beaufighter J-Johnny which Flying Officer Jack Cox belly-landed at Langham after the anti-shipping strike on 29 June 1944. Pictured with the plane are 455 Squadron ground crew, left to right: Corporal Jack Fisher, Corporal Alan 'Tex' Carter and Corporal Ron Badger. The aircraft had been hit by flak in both wings and had lost most instruments. This photograph shows the large supercharger air intakes on top of each engine nacelle and, on the wing next to the engine, the oil cooler intake.

persisted in his attempts to free the pilot until eventually the aircraft sank beneath him.'[35] Although he was only twenty-five kilometres off the enemy coast, Vic Pearson was kept company by relays of Beaufighters from his Squadron led by Squadron Leader Colin Milson, by three aircraft from the New Zealand Squadron, and by Thunderbolts from the United States Army. The air-sea rescue launch finally arrived at Great Yarmouth at 5 pm, some eleven hours after the ditching. Pearson went on the standard two weeks 'survivor's leave' then reported back to his Squadron. He went on to navigate for another eight 455 Squadron pilots in 455 Squadron before the end of the War. For this and later work in 455 Squadron, Vic Pearson was awarded the Distinguished Flying Cross. About this episode the Recommendation reads,

> Despite this harrowing ordeal he returned to his Unit and was on operations within six weeks. Rather than having an adverse effect upon this navigator, his experience made him keener to engage the enemy.[36]

The attack left one minesweeper burning and scored many hits on other vessels.

Ted Collaery's wife of two years, Alice Collaery, was sent a telegram informing her that her husband was missing and believed to have lost his life. The day after the attack, Jack Davenport wrote to Alice, informing her that her husband 'was observed to ditch, and, although aircraft remained in the vicinity for some time and further aircraft carried out an intensive search, as well as an Air Sea Rescue Launch recovering the aircraft's dinghy, he was not seen and is believed to have lost his life when the aircraft ditched'. Davenport added a note about Ted Collaery, 'His thoroughness, keenness and ability

'Collaery was fighting to be freed': Flying Officer Ted Collaery from Wollongong, New South Wales. Collaery was killed on 29 June 1944 after his Beaufighter was hit by flak during an attack on shipping off the Hook of Holland. Collaery successfully ditched his aircraft but he drowned when it sank and he was unable to escape from the cockpit. (Photograph Courtesy B. Collaery)

made him a very valuable member of the Squadron and earned everybody's respect. We shall all miss him very much indeed'.[37]

Alice Collaery received many letters of support at her home in West Maidstone, from friends of hers and from Ted's fellow airmen; many letters expressed hope that Ted would have been picked up by fishermen, or even by a U-boat. This support led Alice to continue to hope that Ted might be alive and she sought help from the Red Cross and the Vatican, hoping for news. An official response in August 1944, to enquiries to the Australian Red Cross stated that, 'it would appear that it would be impossible for your husband to have been found by an enemy vessel without such enemy vessel having been seen in the neighbourhood, moreover, when the dinghy inflated it was seen this officer was not in the dinghy.'[38] Finally, in March 1945, Overseas Headquarters of the RAAF in London advised Alice that, in 'the absence of any news indicating that he may still be alive it is now proposed to take action to presume, for official purposes, that he has lost his life'. Nevertheless, Alice always wondered whether Ted would some day turn up. Ted Collaery's name is commemorated in England at the Runnymede Memorial to Missing Airmen.

Training accidents continued to be a drain on crews. On 1 July, three crews took off from Langham for formation practice. After about an hour of local flying, Flying Officer John Billing from Mildura, Victoria, crashed his Beaufighter near Clay, some eight kilometres from base. One engine had cut out and the Beaufighter spun into the ground. Both Billing and his navigator, Flying Officer Terry Edwards from Auchenflower, Queensland, were killed. Twenty-eight year old Billing and his twenty year old navigator Edwards had arrived at the Squadron only two days before. They were buried at the RAF Regional Cemetery in Cambridge.

Several days later, recovered from his wounding off the Hook of Holland on 29 June, Jack Cox was called for operations again. He recalls,

On the afternoon of July 3rd I was sleeping soundly in an arm chair in the mess. Rain was simply bucketing down and the bar was doing a roaring trade. I was shaken awake by the Duty Officer and asked if I was sober enough to fly. Apart from being a non-drinker I thought he must be joking. He arranged transport to the operations room where Allan Ibbotson was waiting for me. We were to carry out 'Recco No 8' along the Frisian Islands to Borkum then north around Heligoland and back. We had been to Borkum before so briefing was a formality except the Met. forecast was for 'clearing weather'. For take-off visibility was nil. I had practised instrumental takeoff in the Link trainer but this was going to be the real thing.

I lined up on the runway as best I could and and set up the gyro compass on 'zero', and opened the throttles. With the aid of the gyro and the artificial horizon we stayed on the runway and became airborne. At 1000 feet we turned and set course for Borkum. Little did I realise I wasn't to see outside the cockpit for over 2 hours. If anything the rain got heavier and approaching the Heligoland area we ran into a severe electrical storm. Shortly after turning for home, Ibby called me up on the intercom and apologetically informed me that he had only just been able to regain contact with base because of the static. After decoding the message he was much more concerned. 'Unless we had returned to base half an hour ago it would be closed to flying.' No diversion had been offered so back to Langham we went. On arrival we flew over the base at 1000 feet and informed control we were in 10/10 cloud and could see nothing.

They sent us out over the Wash to come down to 500 [feet] and a return pass over the drome revealed nothing. Back over the Wash and down to 200 ft. Control would not believe that we still couldn't see a thing despite their use of vertical searchlights. Somebody then had a brainwave and sent us up to 1500 feet. Before reaching that altitude we burst from the cloud into bright moonlight. At least I could now fly visually and admire the picture of the moon shining on the continuous bank of cloud below us.

Better news was to follow. 'One fighter station is still operating in the South of England near the mouth of the Thames.' (Bradwell Bay) Control told us not to worry about navigation as Fighter Command radar would vector us to the drome. When told we were overhead we still could not see the ground through 10/10 cloud. A searchlight was then used to find a gap in the clouds and on seeing it we dived through and landed. On the next day, 4th July, a 30 min flight saw us back at Langham.[39]

4

Rockbeau Squadron

Within a month, rockets were to be standard armament and 455 Squadron regarded itself as a 'Rockbeau' Squadron.

John Lawson, former 455 Squadron Adjutant.

Crews saw a cargo liner of about 5000 tons completely dissolve in a terrific explosion which could be heard above the noise of their own aircraft.

455 Squadron Operations Record Book, 15 July 1944

For the fighting men there was nothing for it but to fight on.

Grand Admiral Doenitz, former Commander-in-Chief of the German Navy.

IN 1943, rockets were one of the new weapons in the Anti-Shipping Tactics Directorate of the British Air Ministry. The rocket was nearly two metres long, with a main body of diameter close to eight centimetres which was filled with a propellant of cordite. The early rockets had a sixty-pound explosive head containing TNT or Amatol and on the tail were four stabilising fins. This rocket was used initially as an anti-flak weapon.

Coastal Command's first operations using rockets were in June 1943, flown by 16 Group's North Coates Wing against shipping on the Dutch coast. First impressions were unfavourable; they found that synchronisation between cannon and the rocket was virtually impossible with the sixty-pound head because the trajectory of the rocket was much steeper than the trajectory of the cannon. Pilots had to pull up the nose of the aircraft before firing, which mitigated against accurate aiming. As well, the sixty-pound head created so much wind resistance that the propellant could not hold the rocket on a

Beaufighter NE 543 firing a salvo of twenty-five pound rockets in a demonstration flight over The Wash off the coast of Norfolk. This photograph also shows the dihedral tailplanes, swept up to improve directional stability and the .303 inch Vickers machine gun in the navigator's position. Beaufighter NE 543 was delivered to 455 Squadron on 14 August 1944. This aircraft was flown by Flying Officer Bill Stanley, with his navigator Flying Officer Ken Dempsey.

straight course beyond about one thousand metres. After only six operations the North Coates Wing reverted to the use of cannon for anti-flak defence.

In July, the sixty-pound explosive head was abandoned in favour of a more streamlined twenty-five pound solid head, and soon the improvement became apparent. They used a spaced salvo; the four rockets from each wing would leave one at a time, but close together, forming a virtual solid rod. By December, as many as nine ships were sunk by 16 and 18 Groups using the twenty-five pound head. The nose profile of the twenty-five pound rocket was further modified to reduce drag and to give it greater underwater travel, now increased from 22 metres to 55 metres. It was called the 'J-type solid head'.

By 1944, the Germans had begun to correct for the effect of increasing anti-flak aircraft. They were sailing at night to reduce the window of time that the strike wings had to attack the merchant shipping. They were increasing the number of escorts, up three escorts to each merchant ship, and they were sailing close to the protection of shore-based flak batteries. Now, the strike wings moved ahead again, as they all began to use the new rocket. Jack Cox describes the weapon,

> Four rockets were attached on rails under each wing and retained by a piece of low tensile wire. On pressing the 'fire' button an electric current ignited the cordite and it quickly built up a 'tugging' force sufficient to shear the wire.[1]

Aircrew were initially a little apprehensive about using rockets. They were worried about the effect that carrying and firing rockets might have on the

'Four rockets were attached on rails under each wing': 455 Squadron groundcrew loading twenty-five pound solid head rockets on to a Beaufighter TFX. This aircraft (NT 892) arrived at Langham early in August 1944 and was burned after a crash-landing at Langham on 9 September 1944.

characteristics of the aircraft. 455 Squadron was a low-level squadron – very low-level as far as the crews were concerned. The manoeuvrability of their aircraft was essential to them and anything that would interfere with that at all was a concern. During the development of the rockets two test pilots had been killed and it was believed that the rockets were getting 'hung up' on the rails under the wings. Wing Commander Jack Davenport took the lead again. Without having flown with rockets before he loaded them on his Beaufighter and demonstrated that the aircraft had lost none of its manoeuvrability or handling. Coastal Command headquarters heard of Davenport's flight and asked him to help with some experimental work. Jack says,

> The Air Ministry allocated a ship of about 100 tons as a target and anchored it in 'The Wash' off the north Norfolk coast. It was painted in black and white squares for reference. We would attack and sink the ship with rockets and when the tide went out the ship would be welded up and refloated with a different cargo and weight distribution and we would attack it again. I must have sunk that one ship 24 times. We were trying to work out the best tactics and weapons settings. On occasions the 'boffins' would come up with me to observe. I'd ask them which square they wanted hit. They'd tell me, so I'd then ask what part of that square, so they'd decide. 'You'll be bloody lucky if we hit the ship at all' was my response.[2]

Work by the Air Ministry's Operational Research Section concluded that hits in two holds were necessary to sink a ship of 3000 tons. Accordingly, the optimum number of aircraft to attack each ship was planned to be six.

After experimentation, a method of synchronizing cannon and rocket was perfected. At 1000 metres from a target, the aircraft sight would coincide with the fall of shot for the cannons. At 600 metres the sight would mark the fall of shot for the rockets. When a Beaufighter pilot fired his cannons at the superstructure of a ship, at 600 metres, the rockets should be striking the waterline. Jack Davenport says, 'Fired from 900-600 yards a salvo of eight rockets would penetrate the ship's hull, go right through its cargo and structure and come out the far side with a hole of about three feet in diameter'.[3]

In early July, 455 Squadron was working hard. For four days in a row, in over thirty-one sorties, they continued to patrol. One of the sorties left Langham at 3.30 am. These first-light strikes required the crews to be ready for briefing at about 2 am in the morning. A Squadron navigator recalled the routine,

> We'd have to get up and assemble in the briefing hut at about two in the morning. We'd get all the information we needed except the target and then wait for the reconnaissance plane to come back. If the weather was no good the reconnaissance aircraft wouldn't be able to find any targets and the strike would be cancelled. In winter we'd sometimes go to six or seven briefings before we got off.[4]

455 Squadron used rockets on operations for the first time on 6 July 1944. That morning Jack Davenport led an armed reconnaissance of twenty-one Beaufighters to the Frisian Islands. They entered an area covered by dense sea fog and crews saw many balloons protruding just above the fog, showing the location of a large convoy. They wondered if the commander of the convoy knew that he'd given his position away in such spectacular fashion. Davenport flew on down his patrol route, but the fog and cloud seemed to be increasing so he turned the formation around to see if he could attack the convoy under the balloons. There was no practical way of doing this, so Davenport was forced to return to Langham.

That afternoon, forty-three Beaufighter crews from both Langham and Strubby, ten from 455, were briefed for another attempt. The leader of the 455 section was Squadron Leader Lloyd Wiggins. Three of the 455 Squadron crews (captained by Whishaw, Barbour and Kimpton) had already been with Davenport on the morning patrol. David Whishaw recalled,

> Mid-afternoon July 6, 1944, was warm and still at Langham. I remember sitting next to Billy Barbour on the 'B' Flight floor as we waited for take-off. Our briefing had told of a large and heavily escorted convoy off Borkum...It was to be a joint strike operation...two squadrons of Flakbeaus followed in by two of Torbeaus, 455 led by Lloyd Wiggins on the port side of the North Coates aircraft. I was instructed to lead a Vic of five on his port side.
> Billy was a short, slight, cheerful young man and a good and experienced pilot; everyone liked Billy. As we waited we talked about the morning flight and suddenly he turned to me, shook his head slightly and said quietly, 'I don't want

to go on this one', an uncharacteristic remark from him. I didn't know what to say, so said nothing.[5]

The two Langham squadrons were airborne at 7.10 pm. They linked up with the two Strubby squadrons and after they passed over the town of Cromer on the Norfolk coast, the formation of forty-three Beaufighters dropped down to just above sea level. David Whishaw positioned his Vic about 200 metres to the port and a little behind his leader, Lloyd Wiggins, with his own formation. They patrolled to a position off the Island of Nordernay in the Frisian group. David Whishaw recalls,

> In what seemed a very short time the leading escorts of the convoy came into sight on our port bow, steaming west-south-west in light haze. The Flakbeaus of both squadrons at once began a climbing turn to starboard and then eventually to port to bring the whole formation perfectly positioned at right angles to the convoy about two thousand yards away at a height of some fifteen hundred feet. Also positioned behind and well below were the Torbeaus at their correct run-in altitude.[6]

The convoy contained ten merchant ships in two columns, led by three minesweepers and with more escorts on each flank. Following in formation behind Wiggins, the aircraft on the inside of the turn had to fly slowly to keep station, according to David Whishaw, 'hanging on their props'. This had been a well calculated manoeuvre by the leader; he had assessed that he could attack quickly from the shore side of the convoy without being in danger from the shore-based flak batteries. To have attacked from the seaward side of the convoy would have meant clumsy and time-consuming manoeuvring. David Whishaw remembers the convoy defences as the squadrons came into line,

> Almost simultaneously the sky was littered with ugly black explosions of heavy flak from the convoy, joined soon by white puffs of the lighter variety and vivid lines of tracer fire till it seemed that every plane must surely be hit. Next came the now familiar command 'Attack! Attack! Attack!'.[7]

Pilots did their final drills; gun sight swung into position, cine camera on, throttles wide open and into a shallow dive; each pilot to his own target. The flak defences had been light at the start but built up quickly. David Whishaw had in front of him several M-Class minesweepers,

> I selected the third or fourth [minesweeper] to give an unimpeded run-in to the other ships for the two aircraft in my port echelon, opening fire at about a thousand yards, cannon shells first hitting the water then creeping up to the vessel's hull, raising contact to the superstructure and then a touch of rudder each way to rake the length of the deck; pulling out at the last instant with a swerve to avoid the ship's balloon cable.[8]

David Whishaw recalled the incredible noise of engines at full throttle, the rattling of the ammunition feed, the thumping of the cannons in the fuselage and the smell of cordite smoke filling the cockpit. He found that these things

'Luck plays a most important part': Pilot Officer (later Squadron Leader)
Wally Kimpton from Perth, who flew Beaufighters with 455 Squadron from
January 1944 until February 1945. Wally Kimpton commanded B Flight,
and was awarded the Distinguished Flying Cross in October 1944 for his part
in a large number of anti-shipping strikes with 455 Squadron.

seemed to dispel fear. He also heard a crew in trouble – it was Flight Sergeant
Jack Costello from Homebush, New South Wales. Whishaw continued,

> As we climbed away, my navigator, Jack Belfield, spoke urgently over our
> intercom 'there's one of ours in trouble' (Costello) and almost immediately a
> voice broke radio silence in rising and indescribable tension – 'I've an engine on
> fire' A few seconds later – 'I'm going to ditch', and then, moments after – 'I'm
> ditching'. Now came the voice of Wiggins, firmly – 'make it a good one'.[9]

Flying Officer Lee Turner was navigating for Flying Officer Steve Sykes.
After the attack Turner said, 'I was taking pictures when I saw a torpedo hit
one of the merchantmen. There was a pause for a fraction of a second, then
a great mushroom of smoke and water shot about a hundred feet into the
air...The sky seemed to be full of Beaufighters all milling around trying to get
in at the convoy'.[10]

455 Squadron crews also saw Bill Barbour's plane fall into the sea during

'Crews saw Bill Barbour's plane fall into the sea': Flying Officer Barbour from Eagle Junction, Queensland, was shot down by flak near the German coast during a strike against a convoy on 6 July 1944.

the attack, with the aircraft's tail shot off. No trace of this aircraft was found. Barbour had completed sixteen operational sorties with 455. His navigator, Flying Officer Fred Dodd RAF, had joined the Squadron in December 1943 and had completed eighteen operational sorties.

Meanwhile, Flight Sergeant Costello's Beaufighter had hit the water and it broke up. Crews saw the starboard wing and engine break off in the water. The Beaufighter sank almost immediately, but the dinghy came to the surface and a crew member was seen climbing in. Costello went down with the aircraft. It was his seventeenth operational sortie. Next morning, an air-sea rescue boat found an empty dinghy and saw an enemy vessel heading towards the coast. Costello's navigator, Flight Sergeant Robert Taylor, was picked up by a German patrol boat and he survived as a prisoner of war.

Flying Officer Wally Kimpton was also hit. He says,

At first it seemed as though it was going to be easy as we met very little flak. Then it started. Instead of firing up at us as we came in they put up a barrage around the convoy, forcing us to dive through it. Having fired my cannon and RPs

71

[Rocket Projectiles] I finished low to the water and...realized my starboard engine had been hit and [was] not functioning. Before I was able to take any corrective procedures I saw out of the corner of my eye one of our planes ditching. Time was taken up on the trip back to base in feathering, slowly gaining height, trimming and hoping I had manipulated the petrol cross feed mechanism correctly...[11]

Wing Commander Jack Davenport had the duty of writing to the next of kin of Squadron members lost on operations. Fred Dodd's sister Molly wrote back to Jack Davenport seeking hope that her brother might have been thrown clear of the aircraft, rescued by the Germans and taken prisoner. Jack had to reply to her that there was very little chance of him being thrown clear, and that several Squadron crews had remained in the vicinity of the crash, searching, and nothing was found. Bill Barbour's loss was also keenly felt. Doc Watson wrote in his diary,

> Thurs 6th July:
> Billy Barbour (Fred Dodd) had tail shot off on way in. Costello (Bob Taylor) had eng on fire and ditched. Taylor made the dinghy - picked up by Jerry. Cossy probably hit gun sight. Very hard blow to lose Billy - first crew out of B flight...Four pilots from our hut now. Not at all rattled myself but it is hard to see them go.'[12]

The 455 Squadron Adjutant, John lawson, later wrote that within a month of this first rocket attack, rockets where normal and standard armament, and 455 regarded itself as a 'Rockbeau' squadron.

After a month of fighting, the Normandy bridgehead had still not been turned into the spearhead that had been envisaged in the planning. On 7th July, RAF bombers dropped 2,500 tons of bombs on Caen in preparation for an attempt to capture the town. For the Germans, the final message in Hitler's directive on 8 July was that every square kilometre must be defended tenaciously.

455 Squadron was on operations continuously for twenty hours on 8 July. Still searching for German E-boats, the first patrols were away at 3.40 am, the last ended just before midnight. Eleven separate sorties were flown. One of the patrols, by Flying Officer Dave Whishaw, was the fourth of six individual sorties that he flew during his tour. He has written that, 'There is a subtle difference between entering enemy territory alone and when buoyed up in the company of other crews. The tacitly acknowledged "It won't happen to me" syndrome was apt to vanish in these solo sorties'. Whishaw entered the patrol area near the Heligoland Bight, and he recalls that,

> We had climbed to a few hundred feet and immediately noticed a quantity of flotsam in the water. I completed a circuit to starboard and flew low, throttling back as far as was safe. Soon after identifying items of wreckage, we flew directly and unexpectedly over a corpse; a fleeting but vivid glimpse of the body, head back, sharp gaunt jaw agape, torso bare or maybe clothed in a light coloured life jacket. Climbing again we saw on the starboard side a motor boat, later identified

as a [German] 'R' boat travelling north at speed, unnoticed by us during the circuit just completed.

Here was a dilemma. We had been briefed to look for more important targets and our patrol was not yet complete. Conferring on intercom, we decided on a very wide sweeping circle anti-clockwise, to bring us in view of the Frisian Islands and back on reverse course, hopefully to intercept the 'R' boat again. This is exactly what happened and we bore down on the craft in a shallow dive without receiving any fire. As I pressed the gun button a single machine gun returned with a few tracer but stopped almost immediately. At once I had the feeling this was a cold blooded and cowardly thing to do and as we pulled away after what was really a half-hearted strafe, I was also struck with the appalling thought that perhaps Costello's navigator, Taylor had been picked up and was on board. I must have hit something explosive or inflammable with a few shots, because we left the craft smoking and slowed down...We set course for Langham.

Although Taylor had been sighted alive in his dinghy, it was generally thought he had not been rescued and it was a great relief to me to hear years later, that he had survived as a prisoner of war. When on leave with Jack (Belfield, navigator) at his home in Nottingham some weeks later and discussing this trip, it is strange we found we had mutual feelings about the whole mission.[13]

Unexplained losses of crews continued. On 10 July, two Beaufighters left Langham at 1.30 pm on a special reconnaissance along the Danish Coast. One was flown by Pilot Officer John 'Tiger' Payne, with navigator Flight Sergeant Jock Rennie, both back on operations after their harrowing flight on 11 June, when Rennie had bailed out while Payne had wrestled his damaged Beaufighter back to Langham. Just seven minutes after take off, when they were some five kilometres off the Norfolk coast, Rennie saw the plane on their port side make a turn to starboard, 'rapidly losing height as though making an attempt to R/Base [return to base]'.[14] In the plane were Flight Sergeant Max Roberts from Williamstown, Victoria, and his navigator, Flight Sergeant Jack Andrew of Bendigo, Victoria. Payne followed Roberts to find out what was happening. When they were at about one hundred feet the starboard wing of Roberts' plane dropped, the plane hit the sea and sank immediately. Payne circled for fifteen minutes and a light naval craft searched the area. They found nothing except wreckage and an upturned dinghy (each Beaufighter carried a dinghy in the wing and the dinghy was released when an aircraft ditched). Payne headed back to Langham. He was up again within hours to return to the Danish coast with another crew in escort. They sighted numerous fishing vessels, sent a negative sighting, weather and sea reports, then called it a day.

That day, Squadron Leader Milson was awarded a Bar to his Distinguished Flying Cross for leading 455 Squadron with great dash in the combined wing strike on 15 June, and Flight Sergeant Payne was awarded the Distinguished Flying Medal for his skill and bravery in saving his aircraft on 11 June.

The Langham and North Coates wings paid a surprise visit to the Norwegian coast on 15 July. From Norfolk they flew north in good conditions but as they approached Norway the weather deteriorated and by the time they made landfall at Lister, near the southern tip of Norway, they were forced down to sea level. Wing Commander Davenport led with twelve aircraft from 455 Squadron armed with cannon and rockets. 144 Squadron RAF and 404 Squadron RCAF flew the flanks. In all there were thirty-four aircraft to provide anti-flak protection to twelve aircraft of 489 Squadron RNZAF following, as usual, in their torpedo-equipped aircraft.

Davenport turned the formation to fly east along the coast towards Mandal. They came up behind a convoy steaming close to the low islands off the coast, the merchant ships in two columns closely surrounded by their escorts. The escorts were camouflaged and one had an aircraft shape painted on the bridge. Davenport swept in across the convoy, strafing two ships, a merchantman and its escort. As Davenport's cannon shells began to strike the merchant ship, the escort vessel fired rockets which trailed steel wires.

Davenport later gave his account of this for the BBC and his interview, made on a 33 rpm record, still survives.

The weather on the way out was appalling but even though visibility was sometimes less than a thousand yards in heavy rain, we managed to keep formation and make our land-fall exactly on time. A few minutes before we sighted the convoy, the weather lifted making

'The German gunners were...blown off the ships': 20 millimetre cannon rakes the deck of an armed trawler escorting a merchant convoy off the southern coast of Norway on 15 July 1944. Forty-six Beaufighters attacked the convoy of eight ships. This escort ship had the shape of an aircraft painted on the bridge. (Photograph Australian War Memorial)

the attack possible. The convoy consisting of nine ships, was only about a mile off-shore and under cover of land batteries. The merchant vessels were escorted by heavily armed trawlers.

I gave the order to attack and the flak-busters climbed up to about fifteen hundred feet and drew ahead of the torpedo aircraft which remained close to the water. Just before we dived to attack the trawlers opened fire. The flak was quite heavy to begin with, but that was soon silenced. We dived down to deck level firing as we went, and even before we pulled away the largest merchant vessel had already been set on fire by our cannon. During our attack the German gunners were in many cases literally blown off the ships.

In a matter of seconds the torpedo carrying aircraft struck. As I broke away I saw the largest vessel in the convoy was a mass of flames from stem to stern. Just in front of it there came a terrific explosion and steam and water spouted up to three or four hundred feet. When it subsided the ship that had been there was there no longer. She blew up without leaving a single trace.[15]

All the ships were left on fire and some of their crews were seen leaping from the gun platforms. Crews saw a cargo liner of about 5,000 tons 'completely dissolve in a terrific explosion which could be heard above the noise of their own aircraft'.[16] Ted Watson noted in his diary a 'chap in [a] rowing boat going like mad for shore'.[17] The Coastal Command Review noted, 'The

'We dived down to deck level firing as we went': This photograph was taken from Beaufighter Q-Queenie flown by Flying Officer Ted Watson of 455 Squadron during an attack on a German convoy off the southern tip of Norway on 15 July 1944. Twelve torpedo-carrying Beaufighters were escorted by thirty-four Beaufighters providing anti-flak protection. (Photo courtesy S. Milson)

northern shipping routes had been unmolested by our strike aircraft for a considerable time and it seems that the enemy were caught unprepared.'[18]

Heligoland, meaning 'Holy Land', is a small rocky island in the North Sea some forty-five kilometres off the mouth of Germany's Elbe River. The Island was ceded by Britain to Germany in 1890 in return for Germany's recognition of the British protectorate of Zanzibar. It rises to a height of seventy metres at one end and slopes to a sand spit at the other. During the First World War, Heligoland was used as a submarine base by Germany, safeguarding the excursions by the German Fleet into the North Sea. After the War the fortifications were demolished in accordance with the Treaty of Versailles. Heligoland was refortified by Nazi Germany before the Second World War began and the RAF began attacking the island in the first days of the War. Aircrew understood it to be a favoured area for the Germans to assemble convoys.

On 18 July, the North Coates and Strubby wings took on a heavily defended convoy near Heligoland. They attacked with rockets and torpedoes, in poor visibility, and severely damaged merchant ships and escorts but at a cost of two Beaufighters. Only a few days later, on the evening of 21 July, a combined North Coates and Langham Beaufighter wing also struck near Heligoland. This time conditions were good. The Squadron Operations Record Book describes '10/10ths blue' with slight haze. Eleven Australian Beaufighters joined a total force of twenty-six Beaufighters from 144 Squadron RAF and 404 Squadron RCAF (temporarily stationed at Strubby) and 455 and 489 Squadrons. Squadron Leader Colin Milson was chosen to lead. Amongst his Australian crew men were Flight Lieutenant John Pilcher DFC and a thirty year old Beaufighter pilot, Flying Officer Lloyd Farr from Blayney, New South Wales. After school in Sydney Farr had worked in the office of a chartered accountant for six years, and enlisted in the Air Force in 1941 aged twenty-eight years. Farr's aircrew training was typical; initial training, elementary flying training, service flying training in Canada, general reconnaissance training, to an operational unit, and lastly torpedo training in Scotland. Farr finally joined 455 Squadron in October 1943. After only a single operational sortie in Hampdens, he converted to the Bristol Beaufighter.

Twenty-five kilometres south of Heligoland they found the largest enemy convoy to be attacked by Coastal Command so far. Squadron Leader Milson made a recording for the BBC on the day after the attack. He told this story,

They were a hell of a long way off when we first saw them, it must have been fifteen miles away and although it was only about three minutes 'til we were onto them, it seemed an interminable time. They looked rather like a small invasion fleet, and there were actually forty ships altogether. Only nine of them were merchantmen, but these had thirty-one escort ships to look after them. They must have been expecting trouble...

I gave the signal to attack and all of us who were flak-busting dived down into the convoy, shooting at everything we got in our sights. We Aussies were in the centre with the New Zealanders on our port and the Canadians on our starboard side. The RAF squadron flew in at deck level with their tin fish...

76

'Looked rather like a small invasion fleet': Flying Officer Lloyd Farr attacks an escort vessel near Heligoland on 21 July 1944, in the face of flak, balloons and rockets.

> I went in at one of the escort vessels and got in a burst at the depth charges on the deck. One of them went off in a terrific explosion that appeared to blow the stern right off the ship. In ten minutes, a pall of black smoke covered the convoy and practically every single one of the forty ships was either sinking or on fire.[19]

As the other pilots followed in behind Milson, the escorts were becoming enveloped in smoke. Flying Officer Lloyd Farr wheeled in on an escort which fired a rocket trailing a wire as he approached. He had to avoid this, and also the balloon cable that the ship was flying. But then he took a hit on the leading edge of his starboard wing that cut his aileron control. Farr struggled back to Langham and landed safely. Flight Lieutenant John Pilcher's Beaufighter was hit in the starboard engine and it failed almost immediately. Pilcher began losing height despite having jettisoned all movable equipment and he advised that he was preparing to ditch while he still had petrol, but after a few minutes at wave height he found that he could maintain height and decided to try for base. As he flew back towards Langham the air-sea rescue organization was ordered to stand by. With co-operation from the station staff, he flew up a small valley through the shore cliffs and landed successfully. Both Flying Officer Leo Kempson and Flying Officer Colin Cock of Northam, Western Australia, 'pranged on landing' with their tyres shot up.[20] One 489 Squadron aircraft brought back a long length of steel cable wrapped around the wing from a rocket fired by one of the ships.

The combined strike wings had done well, despite heavy opposition. The ratio of escorts to merchant vessels was over three to one. Two merchant vessels were left sinking, two others were left ablaze, five escorts were burning and most others were damaged. David Whishaw says that, for him,

> it was the most satisfactory action of my career. Led by Milson with great skill, the Flakbeaus of 455 were perfectly positioned for attack and in Beaufighter 'F' I was able, as were others, to make the classic shallow dive, throttles wide open, maintaining continuous and effective cannon fire into the superstructure of an 'M' class minesweeper, banking hard to starboard at the very last instant to avoid the vessel's balloon cable.[21]

Colin Milson had had a good day. That night his description of the attack was recorded in the Squadron Line Book. The Line Book had been started back at the station at Wigsley, in the Bomber Command days. It went with them to each station. Bragging about operations was called 'Shooting a Line' and offender's remarks were recorded. Milson was quoted in the Line Book that day, 'The flak was so intense I decided to ignore the light stuff and

Aircrew and Groundcrew together at Langham. Left to right: LAC Ron Castell (instrument maker), Sgt Jack Lefoe (aircraft handler), F/O Jack Belfield RAF (navigator), F/O Bob McColl (pilot), S/Ldr Colin Milson DFC (pilot), F/Lt Jeff Jeffreys DFC (navigator), F/O David Whishaw (pilot).

concentrate on the heavy, though I positioned myself so that the light stuff would enter the aircraft in non-vital spots.'[22] The entry was made by David Whishaw and witnessed by Bob McColl.

Next day, 22 July, this message arrived at RAF Strubby, 'We, the Langham Wing, send our heartiest congratulations to our colleagues of the Strubby Wing on their successful Torpedo and R.P. attack in co-operation with us against the largest enemy merchant convoy blitzed in the War.' The reply sent was, 'Strubby Wing thank you and promise to do even better next time. We go out together.'[23]

30 July was a long and difficult day for aircrew, groundcrew and operations staff at RAF Station Langham. They started early, when twelve 455 and fourteen 489 Squadron crews were woken and briefed for an armed reconnaissance patrol off Norway. Take-off time was set for 9 am, but thick fog was being blown in from the North Sea and the aircraft weren't able to take off until 1.45 pm. Finally airborne above Langham, the crews met their Polish fighter escort and set course for a rendezvous with a squadron from North Coates. On the way to the rendezvous the North Coates squadron lost course in the cloud and returned to their base. So, ten minutes after setting course, the Langham Wing was ordered back to base. The Langham Beaufighters were refuelled in record time and waited to take off again.

Take-off was set twice more, and twice more postponed. They were finally away at 4.15 pm and Wing Commander Jack Davenport assembled the squadrons above the clouds at 1500 feet. Pilot Officer 'Tiger' Payne took off on his third attempt. First he burst a tail wheel which was replaced by groundcrew on the tarmac. Next, he noticed a leak in the engine oil cooler.

455 Squadron Beaufighters line up at RAF Langham. UB-N (Aircraft Number NE 207) was one of the first batch of Beaufighters to be delivered to 455 Squadron in December 1943 and was last flown by 455 Squadron on 24 March 1945, when it was damaged by flak during an attack on shipping at Egersund Harbour in Norway.

He had it inspected by the Engineer Officer and groundcrew. He took off just before the main formation set course for Norway, and he slotted in behind the Beaufighters and in front of the Mustang fighter escort.

On the way to Norway the fog and cloud cleared. Visibility increased to about 25 kilometres. After the effort they put in to get there, they found nothing in the patrol area and turned for home. Just then someone called 'fighters' on the VHF radio and 'a strange aircraft was seen to flash under the formation.' The Operations Record Book records,

> 30 July 1944:
> The Polish Mustang pilots could then be heard chattering volubly on the V.H.F. No one knew what they were saying but there was none-the-less a story in the rat-tat of cannon fire, and now and again a yell of excitement or a cheer as one of the enemy would go crashing into the sea.

At this, according to the Record Book, the Beaufighters 'scorched for home!' The Poles reported twenty-six enemy fighters, and claimed seven shot down. When the Beaufighters reached England the fog was still thick, down to tree-top level in places. The aircraft were diverted to airfields at Coltishall, Rockheath, Horsham and Shipdown. Most landed by 8 pm. That night, the Polish leader rang Jack Davenport and said to him, 'We very pleased thank you so much. We come again yes?'[24] Next day the sea fog was just as thick and it wasn't until 1 August that the crews could fly back to Langham.

Flying Officer David Whishaw was frequently assigned the job of ferrying aircrew around southern England in an Airspeed Oxford or an Avro Anson that were based at Langham. He recalled these trips,

> I had trained on them at SFTS [Service Flying Training School] and was frequently assigned the job of ferry pilot. It was often necessary to take Station personnel to various destination for various reasons. One regular trip was to pick up pilots at Fairoaks aerodrome in Kent, where Beaufighters were sent regularly for major overhauls. More often though, it was flights to London taking ground and aircrews on leave to Hendon Airport under the guise of a navigation exercise. I particularly recall one trip designated a GEE exercise by my principal passenger, our squadron navigation officer Jeff Jeffreys, with official map-reader, Billy Barbour. However, as we were preparing for the journey we seemed to be gathering a lot more hangers-on, the eleventh passenger hareing up to the old crate just as we were leaving the dispersal.
>
> Mostly, an Anson could take off in a few hundred yards. This time, I applied the brakes, got her up to full revs and screeched off, taking most of the Langham runway till she shuddered into the air. With twelve bodies packed in like sardines and a great deal of luggage, it was almost too much for the old dear. Ansons were constructed largely of 'chips and glue' and one could imagine a passenger or two falling through the floor on which most were sitting.
>
> We also had an Airspeed Oxford at times for the ferry job. These were treacherous little aircraft, impossible, we were told, to get out of a spin. In fact, a friend who was an instructor somehow got into this situation, ordered his pupil to bail out saying he would stay with the craft and correct the spin. He didn't; although he was a brilliant pilot. We always limited the passenger load in the Oxford.

On one return trip from Hendon I had on board F/Lt Walker, a quiet and modest man, RAF test pilot for the Bristol Aircraft Co., who was sent to try out a few of the 455 Beaus at Langham. At that time we thought we could throw the Beaus about the sky with some skill and confidence. We were reduced quickly to amateur status watching a master at the controls put the aircraft with unmatched smoothness through a range of manoeuvres. Finally he flew downwind over and past the drome at about 1500 feet, did a half roll and came in to land perfectly at the first few yards of the runway. It was sobering stuff.[25]

July had been a good month for the strike wings and their tactics were working well. That month they had launched seven wing strikes - one off Norway and six off the Dutch coast. From North Coates, Strubby, Langham and Portreath, over five hundred Coastal Command aircraft delivered attacks against shipping. The July issue of the Coastal Command Review noted, 'It must be disconcerting for the German sailor to find that he is likely to be got at practically anywhere on the coast of western Europe.' Coastal Command had benefited from increasing the numbers of anti-flak aircraft in their strike formations and their Review noted that, 'This has been the tendency during July, with the result that the enemy's flak has been swamped with fire and has caused very few losses to our Beaufighters.'[26]

On the evening of 30 July, the German Commander in the West, Field Marshal von Kluge, telegraphed to Hitler's headquarters, 'The whole Western Front has been ripped open...the left flank has collapsed.' Next day, the Chairman of the British Chiefs of Staff Committee, Field Marshal Alan Brooke, wrote in his diary, 'August - the month when wars usually start. I wonder whether this one may look like finishing it instead?'. Now, there were over a million and a half men of the Allied armies in France; Caen had been captured and the American forces were advancing south from Normandy ready to turn east to come around behind the Germans. Unfortunately, the build-up phase of the invasion was running into trouble because of bad weather and because the German E-boat fleet operating from Le Havre, just east of the Normandy invasion beaches, was still interfering with Allied shipping supplying the invasion forces.

As well as the E-boats and destroyers, there were Doenitz's 'third and last contribution to his operations against the invasion, his individual combat weapons'.[27] They were midget submarines and manned torpedoes, similar to those that the Italians had used in Alexandria harbour in 1941. There were a number of types of these weapons. The *Marder* was a one-man electrically propelled carrier for one torpedo. *Biber* was a one-man midget submarine and *Linse* a radio controlled explosive motor boat.

Wing Commander Jack Davenport and Wing Commander Dinsdale, commanding 489 Squadron RNZAF were ordered to the Air Ministry. They were briefed about the situation and Jack Davenport recalls,

The E-boat tactics were to stand off from our convoy protection boats at night and taunt them until one would break from the line and attack them. That was all they needed. The E-boats would wheel around and run away, then sweep in a circle and through the gap in the defences the E-boats would torpedo our convoy ships. We were asked to do something about it. It seemed an almost impossible task. Our attacks had always been done in daylight because we had to dive close to the water. We thought it would take us several weeks to develop the right techniques. They gave us forty eight hours.[28]

Jack Davenport selected six experienced crews to accompany him – Milson, Cock, Kimpton, Thompson, Whishaw and Kempson. The operation was to take place from Thorney Island airfield, on the south coast of England near the Isle of Wight. Aircrew felt that one of the main problems would be gauging height at night. Jack recalls,

> We used radio altimeters but the pilot couldn't watch that when he was in a tight dive at low level. So it would be up to the navigator to tell the pilot what his height was from his normal air pressure altimeter. We experimented with the time and height delay so that the navigator could adjust and read off height. We also practised close formation flying at night. We finally worked out that if we got close enough we could see the glow of the exhaust of the plane in front. In fact we had to tuck right behind the wing of the plane in front. This type of flying was a very over-rated pastime.[29]

The crews flew from Langham to Thorney Island on 1 August, but were forced to wait for several days because of bad weather. The plan was for both Squadrons to fly in tight formation to waiting lines off Cap D'Antifer north-east of Le Havre. 455 Squadron was to patrol eight kilometres off the French coast, 489 Squadron was to be stationed a further five kilometres out and a Wellington from 524 Squadron RAF, carrying flares, was to be five kilometres further out again. A control station operating from an airfield on the Normandy coast was to plot the course of the E-boats as they left Le Havre and pass this information to the Wellington which was to find the E-boats. It would illuminate them with flares to enable the Beaufighters to attack.

They took off in the dark, late in the evening on 4 August. In bad weather they gathered formation over Selsey Bill, a headland south-east of Portsmouth, before setting course for their patrol lines. It was a black night with some cloud cover down to about one thousand feet. The squadrons maintained a tight formation for over two hours, without lights, waiting for the orders to attack. 455 Squadron flew at four and a half thousand feet, the Wellington flew at four thousand feet, and 489 Squadron was at three and a half thousand feet. At one o'clock in the morning, Normandy control passed a message that enemy boats were twelve kilometres from Cap D'Antifer. The Wellington reached the E-boats fifteen minutes later and dropped six flares. Jack Davenport led his squadrons in,

> I led the force of two groups of three planes for these attacks. As leader I didn't have to formate on anyone - it was easier for me, but the others had the difficult task of formating on my aircraft.[30]

The Beaufighters attacked in a dive from four thousand feet, up-moon to the enemy ships. Each dropped bombs from a height of 200 feet on a formation of up to seven small boats amid a barrage of flak. Jack Davenport saw two large white flashes after the attack. The cloud cover over the enemy boats was building during the time that the aircraft made four attacks with cannon. Squadron Leader Milson saw an explosion which he likened to a 'Christmas tree', possibly ammunition exploding. All Squadron aircraft were back on the ground by 2.30 am. Jack Davenport had another unusual experience while they were at Thorney Island. He recalls,

> We didn't attack on successive nights because of the bad weather. One bad night we were waiting in the mess. I was extremely tired - we'd been working for days with almost no sleep and I fell asleep in a chair. Someone put my hat on my head and I felt my foot being kicked. This happened a few times until I eventually got sick of it and lashed out scoring a direct hit. I opened my eyes and saw a uniformed sleeve with one thick stripe and four thin stripes. Half asleep I said 'Jesus Christ'. The sleeve's owner said 'No, but I'd like you to get a new hat'. It was Marshal of the Air Force, Lord Trenchard.[31]

For the Germans, Grand Admiral Doenitz, Commander-in-Chief of the German Navy, summed up the operations of his individual combat weapons against the invasion,

> Handled with courageous determination these weapons, too, succeeded in sinking a number of enemy ships. But against greatly superior enemy forces these few successes scored by our naval craft could not, of course, have had any material effect. Furthermore, enemy domination of the air very quickly made it impossible for our vessels to pick up requisite stores from any base in the invasion area, and it was not long before their activities were brought to a standstill.[32]

For Doenitz the realities became starker and plainer. He wrote after the war,

> Once it [the invasion] had succeeded, however, we found ourselves faced with war on a number of land fronts. The war could no longer be won by force of arms. To make peace was no longer possible since the enemy would have none of it until Germany had been destroyed. For the fighting men there was nothing for it but to fight on.[33]

5

First and Last Light Strikes

There is no doubt that this victory is due to the long and patient hammering of the convoys by direct attack.

RAF Coastal Command Review, August 1944, on the disruption to Germany's import programme.

It was one of our worst days, we lost three good experienced crews.

Warrant Officer Ivor Gordon, 455 Squadron Navigator, on the strike in the Dutch Bight on 10 August 1944.

Under all circumstances we will continue this battle until...one of our damned enemies gets too tired to fight anymore.

Adolf Hitler, to his Generals on 31 August 1944.

PARIS was liberated in August 1944. That month also saw the giant outflanking sweep by the Allied invasion armies in France. In his directive issued on 4 August 1944, General Montgomery, commanding 21 Army Group (the land component of Operation "Overlord") said, 'Everyone must go all out all day and every day. The broad strategy of the Allied Armies is to swing the right flank towards Paris and to force the enemy back to the Seine.' The evening before, Field Marshal von Kluge (commanding the German forces in the west) had asked permission to withdraw, but Hitler replied with a categorical order that von Kluge must launch a counter-offensive. At this stage, von Kluge was anxious to demonstrate his loyalty. The trial of those involved in the plot to assassinate Hitler on 20 July had already begun and Kluge was not prepared to provoke the Fuhrer's wrath by challenging his order, even though he knew it would mean the death of an Army. The historian Chester Wilmot wrote that, 'The forty-eight hours, which began on the afternoon of August 6th, settled the fate of the German Armies in Normandy. When they should have been withdrawing eastwards to the Seine, Hitler drove them westwards to destruction.'[1]

There was worse coming for Germany. During August, following pressure from the Allies, through the intensification of sea mining and because of the increasing scale of attacks on shipping by Coastal Command between the Ems and the Elbe, Sweden stopped insuring its ships trading with German ports. This effectively severed Sweden's trade connections with Germany. It only increased pressure on Germany to have the remaining tonnage get through. The Coastal Command Review for August 1944 noted that, 'There is no doubt that this victory is due to the long and patient hammering of the convoys by direct attack'.

In August 1944, with two of Coastal Command's Beaufighter squadrons detached from their station in Scotland to a temporary station at Strubby in Lincolnshire, seven of the nine strike squadrons were in the North Coates-Langham area.[2]

Squadron Leader Lloyd Wiggins took nine other 455 crews to meet the press at nearby North Coates on Monday 7 August. They stayed there overnight and next morning they were a little surprised to be briefed for a strike off Norway, in company with the North Coates Wing. It was to be a huge force, thirty-nine Beaufighters, with an escort of forty-eight Mustang fighters from both the United States Army Air Force and Polish RAF Squadrons. Twelve of these Mustangs were to escort the Beaufighters; the other thirty-six were planning to strafe a German airfield near Stavanger. Led by Wing Commander Paddy Burns of 254 Squadron RAF, the force took off at 12.30 pm, formed up over the North Coates field and set course for a point on the Norwegian coast just south of the lighthouse at Obrestad, near Stavanger. From North Coates it was over two hours flying to the patrol area and conditions were clear when they arrived. They wheeled south and just past Egero Light found a convoy of fourteen ships, including five to seven merchantmen, the dark shape of the ships silhouetted against the white clouds.

On the run-in to attack they encountered flak from the escorts and shore batteries. A heavy flak shell exploded between two Beaufighters in the 489 Squadron section, blowing a wing from one aircraft and the tail off the other. A 489 Squadron pilot, Warrant Officer Don Tunnicliffe, reported that 'Both fluttered down like two stricken moths, to crash into the sea about 50 yards apart'.[3]

455 concentrated on the escorts again, and six of their Beaufighters were able to take targets. Wiggins took on three – an escort, a minesweeper and a *Sperrbrecher*. Another New Zealand Beaufighter was hit, and crews saw fire streaming from one engine. Don Tunnicliffe caught up with this Beaufighter. He writes,

> It turned out to be Squadron Leader Peter Hughes and Freddie Spink from our 489 Squadron. I pulled up to fly parallel with them but couldn't do a thing to

A low pass by Beaufighters over the RAF Langham Station at Norfolk in August 1944. The Beaufighters were flown by, from left to right: Flying Officer Leo Kempson, Squadron Leader Colin Milson DFC and Bar, and the 455 Squadron Commanding Officer, Wing Commander Jack Davenport DSO, DFC and Bar. (Photo courtesy S. Milson)

help. I don't know what really happened to Freddie Spink as I only glimpsed an object falling from the plane and surmised that Hughes had told Freddie of his intention to bale out so Freddie jettisoned his cupola to do so and had undone his safety harness but had failed to clip on his parachute beforehand. Sadly he would have been sucked out to hit the sea. From 1000 feet it would have been like concrete, rendering him unconscious, to finally drown. Meanwhile Peter Hughes was getting burned about the legs as the flames were sucked back so he got rid of his sliding hood and was simply sucked out to pull the rip cord and float down into the sea, where he finished up being rescued by the Germans and ended the war in a P.O.W. camp.[4]

The attack sank a merchant vessel and damaged four armed trawlers. Three Beaufighters and three Mustangs were lost. 455's Warrant Officer John Ayliffe says that,

I gather that the provision of so many fighters was to tempt German fighters from Stavanger air base, which was only minutes flying time from the convoy. The planning was successful because the Germans took the bait, and there was a decent dog-fight over the convoy although there is no mention of it in the Operations Record. However, the element of surprise was negated because the Polish pilots broke radio silence and nattered to themselves most of the way across the North Sea. They relished any opportunity to have a go at the Germans.[5]

The crews were back at North Coates just after 5 pm that evening, after an operation lasting over four and a half hours.

As difficult as those days were, they are still remembered with affection. Katie Ivatt (a young WAAF at Langham) remembers the Australians, and in particular Flying Officer Bob McColl. She writes,

It was a magical time, the summer of '44. The Australians were great fun and I loved them dearly. I drove for 'B' Flight and went out with Bob occasionally. Just on our bikes round Norfolk lanes in lovely summer weather. A retired Air Commodore who lived in Blakeney gave a dance for the squadrons and I went with Bob. We were allowed to go in civilian clothes. 455 had a big strike on that night [8 August 1944] against German shipping off Norway and he was itching to go. Thank God he didn't: I think we lost six aircraft. There was always this dread though one never mentioned it.[6]

The Official History of the Royal Air Force in the War notes, 'By 9th August all was won and lost in Normandy. American tanks were in Le Mans fifty miles west of the advanced German positions. The Canadians...were moving slowly on Falaise...a pocket had been formed...[containing] the remnants of sixteen German divisions'.[7] The northern side of the Falaise pocket was held by British and Canadians, and the southern side by the Americans.

Also on that day, 9 August 44, Warrant Officer Noel Turner and his navigator Warrant Officer Ted Holmes left Langham in their Beaufighter for a twenty-minute air test. As part of that test Turner stopped one engine and feathered the propellor. He then tried to restart, but couldn't. He says,

Not to worry, I positioned myself with plenty of air space for [a] single-engine landing. Selected wheels and flaps down - no luck, they just hung there in space with no hydraulics operating to send them all the way. Suddenly things became serious, running out of aerodrome and with too much drag to risk going round again on one engine. I said, 'We're going in, Ted' and closed the throttle.[8]

His navigator, Warrant Officer Ted Holmes RAF, strapped and locked himself in (Holmes had previously survived a bad air crash on 1 May 1944 when his pilot, Pilot Officer Neville Wilson, had crashed their Beaufighter at nearby Little Snoring. Wilson had died in that crash). Now, Noel Turner took his plane down. He continues,

In my anxiety to get down before looming obstructions I virtually drove the poor Beau into the ground. We bounced once hard, then stopped. The starboard wing tanks caught fire immediately. I made the customary pilot's obeisance into the gun sight placed there for just that purpose, in the event of a prang or ditching. Most pilots are KO'd by the blow but those with a thick skull say 'Ugh' and commence bleeding freely.

Holmes climbed out of his navigator's cupola and went forward. He helped Noel Turner from the cockpit, which was then surrounded by flames, and the two made their way down the port wing root. They went to the safety of the rear of the plane but got back even further when the cannon shells started exploding. In the sick bay later, their Commanding Officer, Jack Davenport, visited the two men. Noel says he had 'a bad moment' when he recalled it was Davenport's plane that he'd demolished. Noel was then taken to Ely hospital, a leading burns hospital. He says,

I will always be grateful for their care and expertise. My problems (first degree and one second degree burn and ten stitches in the scalp) soon faded to insignificance beside other patients. Those young nurses were the heroines of the piece, coping with pieces of charred flesh that had once been human beings. I don't know how they coped.[9]

Noel says that he was nagged by the thought of getting the fear of flying. He wanted to get out quickly and find out. He continues,

I soon found that importance was placed on whether burns had ceased to turn blue on exposure to the morning air. On the medical superintendents weekly visit I was asked this very question. 'Oh no sir, not any more, hasn't been blue for days' I said. 'That's funny, it's blue now', the superintendent told me. But I was out that very day.[10]

Noel Turner took the mandatory leave, then went back to his Squadron. His next flight was an air-test at dusk. He landed and applied brakes, but had none. He remembers his choices. He could open the throttle, take off and go around again, or hope to come to a stop in time. Noel got away with a bumpy ride over some rough ground at the end of the runway after he decided to ride it out. His remaining legacy of that crash was a surrealistic haircut that earned him the nick-name of 'Herr Kaufman'. Noel says that his Flight Commander 'whacked the trips into me when I returned which I thought was both fair and reasonable - give the blokes who had to double up for me a bit of a break.' For his part in helping Noel from the burning plane, Ted Holmes was later awarded the George Medal. Noel was back on operations on 15 October, and was on four of the next five Squadron operations.

Langham Wing was back in the Dutch shipping lanes on Thursday 10 August. A convoy had been reported off Heligoland and Langham was ordered out - a maximum effort, requiring all available aircraft. This was the day after Tasmanian Flight Sergeant Jim Brock's 21st Birthday. Warrant Officer Allan Ibbotson recalls that, before the operation, Jim had told him that the Wing Commander was going to 'give him a spell from operations because he'd had a fair run, but 16 Group had called for a maximum effort and they [Brock and his pilot Jones] were sent out.'[11] Brock's pilot, Warrant Officer Bill Jones, a drug company employee in Coogee, Sydney, had enlisted in the Air Force in September 1941. He had a badly burned face from a fire in an aircraft crash some months before. That crash had been enough to see him out of the War permanently, but 455 crews say he talked his way back into a squadron. He was called 'Pundit' because a Pundit was an illuminated navigation marker flashing a morse signal. Bill Jones was twenty-eight years old.

Early in the evening of 10 August, Squadron Leader Colin Milson led a formation of twenty-five Beaufighters into the air. Milson took off first, throttled back, flew on straight for several kilometres then circled to port to let the other Beaufighters catch up and gather formation.

Warrant Officer Ivor Gordon was in that formation with his pilot, Flying Officer Clive 'Tommy' Thompson. Gordon was from Perth in Western Australia, and like many others had joined the Air Force from the militia.

Flying Officer Clive Thompson was from Stanmore, New South Wales. He had served on 101 Squadron RAF as a navigator on Wellington bombers. During a flight to the Middle East when his squadron transferred there, Thompson's aircraft had been attacked by enemy fighters and was forced down in Portugal. The crew was interned for nearly three weeks but escaped and made their way to Gibraltar on a small coaster. Thompson was given permission to re-muster as a pilot, and he crewed up with Gordon at their Operation Training Unit. Gordon describes this crewing-up process,

> We did our conversion to Beaufighters at Crosby-on-Eden. During the first few days there we were told to go to the 'Dog and Gun' hotel, to have a few beers and to come back as crews. As I recall there were six navigators and four pilots at the hotel. I thought Bill Roach's pilot was the most promising bit of material, but after a few beers Tommy approached me and I accepted, much to my regret I thought. He'd had a few beers and looked a bit of a mess. But he was the only one of the four pilots to survive the war, so the choice wasn't all that unhappy for me. Clive Gordon Thompson, or Tommy, had done a tour as a navigator on Wellingtons in Bomber Command and as a reward for that work he was allowed to do a pilot's course, even though he was probably not tall enough.[12]

Clive 'Tommy' Thompson and Ivor Gordon had joined 455 Squadron together on 9 May 1944.

As the Squadron circled over Langham, Milson's plane developed serious engine trouble and he dropped out to land. The new leader, Flight Lieutenant John Pilcher, had his navigator fire a Very pistol as he crossed the airfield at 7.45 pm, to signal that he was setting course for the patrol area. Pilcher led ten 455 crews and twelve from 489. They crossed the Norfolk coast at 180 knots and 1500 feet. An hour and a quarter later, at 9 pm, they arrived in the patrol area with Pilcher's anti-flak aircraft at 1000 feet and the New Zealand Torbeaus behind, at sea level. It was hazy as they patrolled the coast for over an hour. They were now south of Heligoland in the twilight, and were just turning for home. Pilcher says, 'There was a thick haze over the sea and I heard one of the torpedo-carriers who were flying below us call out that he had spotted a possible target.'[13] They were just north of the island of Wangerooge. There were five merchant ships surrounded by many escort vessels. Warrant Officer John Ayliffe remembered that sighting,

> [a] huge sea area covered by the convoy and its escort vessels. The outer ring of escort vessels was some miles from the centre of the convoy. It must have been a very important convoy to have such a large number of escorting flak ships.[14]

Pilcher wheeled around to attack and came up against intense flak from the escorts and from shore batteries. Pilot Officer 'Tiger' Payne says, 'Someone ashore was signalling with a lamp to the convoy as we came through the haze but we were on them before they could open up with their flak. When they did though they certainly let fly.'[15] It was one of the heaviest defences they had ever encountered.

The pattern of tactics for strike wings was now established. The anti-flak Beaufighters led in first in a shallow dive firing cannon and rockets, picking out the escort vessels. Flying Officer Jack Cox recalled, 'several of the vessels had balloons flying, and one escort even had two. They were firing fairly steadily but we all pressed home our attack and soon the gunners were ducking for cover.'[16] Four of the Beaufighters were nearly at the end of their run-in, only a few hundred metres from the ships, when they were hit. Warrant Officer Ivor Gordon was navigating for Flying Officer Clive Thompson in the lead section. He was watching out the back of his cupola as Thompson attacked the leading minesweepers with cannon and rockets. Gordon says,

> It was the usual formation, 455 suppressing the flak with the Torbeaus about 1500 yards behind. I saw three [of the 455 Squadron] planes go down in quick succession. It was one of our worst days, we lost three good experienced crews.'[17]

The Torbeaus followed in, scoring three hits on the two larger merchant ships. The New Zealanders also lost a crew who were with the anti-flak sections. Warrant Officer John Ayliffe recalls that, 'one of the ships was ablaze from end to end and the heat was so severe that its outline glowed in the rapidly approaching night.'[18] The aircraft all broke away and made their way back to Langham, some singly, some in small groups in loose formation. The force landed back at base at about 11 pm. The Squadron Operations Record contains this entry,

> 10 August 1944:
> No 455 Squadron Flakbeaux attacked first with cannon and R.P.[Rockets] in the face of the most intense accurate light and heavy flak yet experienced by this Squadron. Unfortunately, as the attack developed, aircraft D, H and L /455 and one aircraft of No. 489 Squadron were shot down. This is the the severest loss yet sustained by the Langham Wing on any attack.

Those crews were Flying Officer Leo Kempson and his navigator Flying Officer Ray Curzon, Warrant Officer Geoff Batchelor and navigator Pilot Officer Morris, and Warrant Officer Bill Jones with navigator, Flight Sergeant Jim Brock. They were all experienced crews; the pilots had an average of twenty-one operational sorties behind them.

Allan Ibbotson later rang Jim Brock's brother Joe, an airman at the Operational Training Unit at East Fortune (Joe later joined 455 Squadron). The body of Flight Sergeant Jim Brock was later recovered from the sea, and he was buried in Hanover War cemetery. Now, after just over four months with 16 Group at Langham, 455 had lost nine crews on operations and another three aircrew in training. This amounted to the loss of nineteen men from the Squadron which had an average of sixty-seven aircrew over that period.

It was another early start for the Langham Wing on Sunday morning, 13 August. They set out in force again for Heligoland, taking off in the dark at 4.56 am. There were Beaufighters from 455, 489, and 254 Squadron RAF at North Coates and an escort of Spitfires. Rendezvous was over a flare-path laid by a Wellington bomber some twenty kilometres off the island of Borkum. Just on dawn they lined up in formation and set course up the

Dutch coast. They flew up by Nordernay and Wangerooge, and off Heligoland, at about 6.30 am, found a convoy of six merchant ships and nine escorts. The merchant ships were in two columns, led by two M-Class minesweepers flying balloons. The Beaufighters attacked and one of the merchant ships and most of the escorts began evasive moves. Cannon shells straddled the leading minesweepers, and an auxiliary on the starboard quarter received a direct hit from a rocket salvo. An armed trawler at the rear of the convoy seemed to be very modern and heavily armed; it was given special attention, being hit by sixteen rockets. After the strike, Beaufighter pilots wheeled for home. Ivor Gordon, navigating for Clive 'Tommy' Thomson says,

> We'd just passed over the convoy and were heading west towards England. I was looking out the back and saw a shell coming straight for us. It must have been an 88 because it seemed quite large. I yelled at Tommy to turn. He did it straight away which was just as well because it burst right where we would have been. They sent another half dozen shells after us and we repeated the exercise several times.[19]

Flight Sergeant Norm Steer from Prospect, South Australia, was hit by flak and crash-landed back at Langham. By 8.20 am, the surviving aircraft were safe on the ground at Langham.

As successful as the Langham and North Coates wings were along the Dutch coast during this period, the highlights of Coastal Command's strike wing activities in August were in the Bay of Biscay. Germany was continuing to attempt to run coastal convoys along the western coast of France to relieve their hard-pressed land communications, and Coastal's Beaufighters, Mosquitoes and Halifaxes combined with the surface forces of the Royal Navy to play havoc with German shipping there. On 14 August, twenty-six aircraft from the two Mosquito squadrons at Portreath combined to attack a destroyer and *Sperrbrecher* at Le Verdon, near Royan, north of Bordeaux. The German naval force put up such a terrific defence that four of the Mosquitoes were lost and five others were damaged. Just minutes after this attack, thirteen Beaufighters from the North Coates and Strubby wings entered Arcachon Bay, just to the south-west of Bordeaux, where they attacked a floating dock, escorts and tugs. Le Verdon was attacked again on 21 August by Portreath, and then three days later by North Coates and Strubby.

Flying Officer Jack Cox had a busy start to August. He was on the strike to Norway on 8 August and again when 455 Squadron went to the area off Heligoland on 10 August. He flew a practice with rockets on 14 August and on the 15th took Beaufighter UB-Y for an air test. Returning to base he lined up the runway and throttled back. At that moment the port engine control cable fractured and with the starboard engine throttled back and the port engine running fast the plane swung to starboard. Jack remembers,

> It would have been dangerous to try to land with one engine on full power so I

quickly retracted my undercarriage and did a belly landing in a cabbage patch next to the airfield. My Squadron Leader, Colin Milson, was quite cross that another Squadron aircraft had been damaged, and he asked for my log book to make a 'red entry'. I went on leave and when I came back I found that the Station OC, Group Captain Clouston, had obtained my log book and made this entry (in green), 'This officer in a difficult situation, displayed noteworthy initiative and skill, and by his cool and prompt action saved both aircraft and crew from possible destruction'. That was probably a better entry than that which Colin Milson would have made.[20]

Flying Officer David Whishaw also had a difficult landing the following day. He recalls,

The ruggedly built Beaufighters usually held together very well in belly landings, though it is doubtful if any survived the damage to fly again. In stark contrast, de Havilland Mosquitoes, though magnificent aircraft in their own right, tended to disintegrate. A substantial number of the 455 aircraft were subjected to the indignity of 'no wheels' landings and mostly were put down with great skill by their pilots, in some instances even when the latter were wounded. The most notorious exponent of this manoeuvre must surely have been 'Flak Jack' Cox who successfully completed a remarkable total of four. It came to be said about Jack that he 'didn't care much whether he came in with his wheels up or down!' I escaped this experience but had my share of single engine landings. These usually did not present problems, but there were variations. Going out on a sortie over the North Sea on 16 Aug in 'U', the aircraft suddenly yawed violently to port with the throttle lever loose in my hand. Initially, I couldn't understand what had happened as the propellor was rotating at about a thousand revs which still gave some assistance. Having realized that a throttle linkage had broken, I trimmed the rudder and turned back to Langham at reduced speed and as was the usual practice, requested and was granted permission for an emergency landing. I made a more or less normal approach and wondered why we were floating along and yawing, this time to starboard. As we eventually touched down, realisation dawned too late that the starboard motor was throttled right off and the port was still doing its thousand revs. All that was needed was to cut the ignition switches, but by this time the aircraft had left the runway on to rough ground and I was fighting to prevent a catastrophic ground-loop. Slowly the aircraft came under control and we bumped our way back to the tarmac without damage. The same thing happened a few weeks later. This time, having made sure of a correct approach I cut both ignition switches at about a hundred feet and glided in without incident.[21]

The SHAEF Intelligence Summary reviewing the situation in the West noted that, 'Two and a half months of bitter fighting, culminating for the Germans in a blood-bath big enough even for their extravagant tastes, have brought the end of the war in Europe within sight, almost within reach. The strength of the German Armies in the West has been shattered, Paris belongs to France again, and the Allied Armies are streaming towards the frontiers of the Reich.'

'The formation wheeled straight in': Flying Officer Forbes Macintyre took this photograph from his navigator's cupola during a strike against a convoy off the Frisian Islands on 25 August 1944. On this afternoon fog and haze made conditions difficult for aircrew. (Photograph courtesy N. Smith)

Coastal Command's strike forces continued to dominate. On 25 August Wing Commander Jack Davenport led his Squadron again. He took a force of forty-four Beaufighters to the Dutch shipping lanes. They left late in the afternoon and patrolled at two hundred feet, sweeping along the shipping lanes off the Frisian Islands. Visibility was down to one kilometre with a thick sea mist. They patrolled for twenty minutes, then Davenport led the formation in a climbing turn to starboard, up to 2000 feet, and finally he ordered a return to base. Some eight minutes later one of the Mustang fighter escorts saw two ships to port. The formation wheeled straight in and attacked immediately. The twelve 455 Beaufighters each attacked with rockets and cannon, diving to as low as 200 feet. Conditions were not easy – bad visibility kept Davenport from using his rockets and manoeuvre room was restricted. Flying Officer Ted Watson was crowded out by another Beaufighter from 489 Squadron. Several escorts flew protective balloons on steel wires up to a height of five hundred feet and Warrant Officer John Ayliffe recalled, 'the low cloud and poor visibility...also it was the first time I had encountered ships in convoy flying balloons. I remember sighting the balloons flying above the clouds but there was no sign of any ships'.[22] The sea mist had been masking a convoy of six to eight merchant ships, flanked by escorts and preceded by two M-Class minesweepers. The attackers scored many cannon strikes, and later photographs showed a dense column of smoke rising to two hundred feet, with smoke coming from at least four places on the after-deck of one of the escorts. All the Langham aircraft were back by 11 pm.

Tuesday 29 August was another evening strike. There was a maximum effort from 455, thirteen Beaufighters in a force of forty-six, escorted by seven American Mustang fighters. The aircraft were signed over from the ground-crew, started, and took off in echelons at one minute intervals from 6.20 pm, led by Sykes, Whishaw and Milson. Milson now had twenty-seven operational sorties with 455, amounting to over seventy-five operational hours. He was well tuned to the dangers of this work.

All aircraft formed up over Langham, and this combined wing was led away by Wing Commander Paddy Burns from 254 Squadron RAF, North Coates. The wing flew towards the Dutch coast at two hundred feet and some twenty minutes later the anti-flak Beaufighters began to climb, leaving the torpedo aircraft at two hundred feet. Soon after they saw a convoy some twelve kilometres away. There were five ships in line ahead, including two large *Sperrbrechers*.

Burns swept straight in. The anti-flak Beaufighters started their attack in a low dive from 1500 metres. Firing first with cannon to range and then at 800 metres firing rockets, they switched back to cannon until about 50 metres before pulling away. Six Australian pilots took on the third escort in line, a *Sperrbrecher,* which was then hit by a torpedo from a Kiwi Beaufighter, left blazing and settling down by the stern. Pilots broke away low to reduce the danger from flak and they left all five ships well on fire. On the other side, the Beaufighters were well damaged by flak which seemed to build up as the strike developed. One New Zealand crew ditched nearby, another Beaufighter was seen heading towards the German coast with a propellor feathered, and Sykes of 455 had minor flak damage. Milson's Beaufighter was also badly damaged, with both engines and the cockpit hit. He was 480 kilometres from Langham with his port engine on fire. He struggled with the starboard, managed to keep it running and made it back to base, flying the the last hour in darkness. For that effort and other operations Colin Milson was appointed a Companion of the Distinguished Service Order. His father sent a telegram of congratulations and later Milson wrote back,

> Your cable regarding that bit of one engine flying in August was rather a shock as I didn't know that the news people at home had spread the story. They shouldn't really as it only worries one's parents - however it was just one of those things that happened and which you must expect at odd times. It didn't worry me in the slightest as I was out on 'ops' again the next day and also the following one. You get pretty hardened to such things and a miss is as good as a mile anyway. I've

always been very lucky and have had a lot of experience on my type of work but for once I became very careless and casual and had my hand well slapped for my trouble.[23]

That day, as the last of his men were crossing the Seine in headlong retreat to the East, Field Marshal Walter Model (Kluge's successor as commander in the west) reported to Hitler that of the sixteen infantry divisions he had got back over the Seine he could raise sufficient men to form four, but was unable to equip them with more than small arms. France was lost. Hitler said at a conference with his Generals on 31 August, 'Since the year 1941 it has been my task not to lose my nerve, under any circumstances; instead, whenever there is a collapse, my task has been to find a way out and a remedy, in order to restore the situation....Under all circumstances we will continue this battle until, as Frederick the Great said, one of our damned enemies gets too tired to fight anymore. We'll fight until we get a peace which secures the life of the German nation for the next fifty or hundred years and which, above all, does not besmirch our honour a second time, as happened in 1918.' It was also on that day, 31 August 1944, that Prime Minister Churchill sent a message to the Air Officer Commanding Coastal Command, which was copied into the 455 Squadron Operational Log: 'I send to you and all your officers and men my congratulations on the splendid work of Coastal Command during the last three months...Most effective attacks have been delivered against enemy shipping, and very many hostile escort vessels and merchant ships have been sent to the bottom'.

For the Allies, the situation in France was getting more complex, and more demanding every day. General Omar Bradley, commander of the First US Army in France, had acute supply difficulties. Eisenhower had given him the task of advancing deep into Belgium but his supply lines were getting longer, his liability larger. He had the additional burden of feeding the French capital. The Americans established the 'Red Ball Express', a circular one-way traffic route between the Normandy bridgehead and Paris. They ran it twenty-four hours a day and only vehicles of the Express were allowed on the route. Eisenhower badly needed ports east of the invasion area to shorten his lines of communications.

British forces crossed into Belgium on 2 September and the next day entered Brussels. This was the fifth anniversary of Britain's declaration of War on Germany and it was on this same day that Hitler issued a directive to his Army commanders. He ordered Boulogne, Dunkirk and Calais to be held.

That afternoon, 3 September, Squadron Leader Colin Milson led a Beaufighter reconnaissance force to the vicinity of Egero Light in southern Norway. It was a very long patrol, five hours flying and without an attack being made. Still, it was a testing trip. There were forty-six Beaufighters from Langham and Strubby, and nine Mustangs in escort. They started their patrol off the Lister lighthouse, at 1500 feet. As usual the leader of the anti-flak force, Milson, led the patrol. They found a cargo liner of about 5000 tons heading due north up the Norwegian coast. Crews saw two promenade decks, a red cross and a red strip on the hull. There were three trawler type auxiliary ships around it and they seemed to be heading towards the coastal port of Egersund.

As the Beaufighters closed to investigate, they came under fire from several shore positions, and soon after from the escorts. Some crews reported flak and rockets fired from the cargo liner which seemed to be a hospital ship, but that was not corroborated. Milson decided not to attack and he turned the Beaufighter formation away, although the shore flak continued to fire at the strike wing a long way out to sea. Flight Sergeant Jack Tucker from Carlton, Victoria, has since said that, 'I have always felt that the flak did come from the liner and I recall that Col Milson seemed to hesitate before calling off the strike. The ship was close in-shore so I could be mistaken and that it could have been from a ground position.'[24] The maximum endurance of the Beaufighter TFX was five and a half hours and on the way home, after nearly five hours on this patrol, some aircraft jettisoned their rockets and one of the New Zealand 489 Squadron Beaufighters jettisoned its torpedo. All Beaufighters made it safely back to base.

From September, Ultra decrypts confirmed what the aircrews had known. The Germans were protecting their convoys more heavily and sailing only by night. The Allied attack against German shipping was comprehensive, carried out by submarine, by the Home Fleet with the carriers *Indefatigable* and *Implacable*, by the Norwegian Motor Torpedo Boat Flotilla, the small craft of the Special Operations Executive unit based in the Shetlands, and by the strike wings of Coastal Command. Coastal Command received intelligence on German defences and convoy routes from photographic reconnaissance, from the Special Intelligence Service and the Special Operations Executive. Coastal Command recorded that, during July and August 1944, only one German merchant convoy was found sailing in daylight off the Frisian Islands.[25]

On the afternoon of Wednesday 6 September 1944, thirteen 455 Squadron crews were on standby. One of 455 Squadron's more experienced pilots, John Ayliffe, describes the preparations for strikes,

> Crews on standby for strikes were notified by the daily MAYFLY. Generally some hours of notice was given when an operation was imminent. Group relied on information of shipping movements radioed by reconnaissance aircraft and other intelligence reports. Once it was decided to mount an operation crews were called to the Operation Room for briefing by the station Tannoy. After an operational meal, crews assembled in the Ops Room. On entering the Ops Room crews were able to see immediately where the strike area was located. This was shown on a large map with the route already shown with bold red lines.
>
> The Station Commander or the Wing Commander of the Squadron would outline what the target was and, if known, the composition of the convoy, the important ships and the escorting force and their location in relation to the merchant vessels. The importance of surprise was stressed viz., low flying to negate the radar, radio silence and the composition of the whole strike force.

The crews were then briefed by the Navigation Officer, Armaments Officer and Meteorology Officer. The Intelligence Officer would then provide information on whether enemy fighters were likely to be present, also if the strike force was to have a fighter escort. Intelligence would also point out hot spots for anti-aircraft fire, and navigational points such as lighthouses, villages, lakes, railways etc.

After all departmental briefings the leader of the strike force would then discuss tactics, the best way to attack the convoy, and in the case of a multiple squadron strike, where each squadron would fly in formation.

Before crews were taken to their aircraft the Navigational Officer would take the navigators aside so they could be instructed on the finer details on their navigational responsibilities, and told who would be the 'lead navigator'. This was generally the Strike Leaders navigator, but each navigator had to keep his own plot. Other information such as codes for any radio used after the attack and the Colour of the day (which was secret information) were given to the navigators. The navigators were also informed of alternative landing points on return in the case of bad weather or fogs.[26]

Squadron Leader Lloyd Wiggins led twenty-six Beaufighters from Langham to their familiar patrol area off the Dutch Frisian Islands. They had six Mustang fighters from Coltishall as an escort. At a quarter past eight that night they caught a convoy forming up and attacked. Their main targets were four merchant vessels in line ahead; all seemed to be over two thousand tons. They attacked in normal fashion; the Australians first with cannon from two thousand metres, then rockets from 800 metres. They took the attack down close to the convoy, to within one hundred metres of the targets before pulling away. There was crowding as the Beaufighters converged on the targets. Flight Sergeant Jack Tucker was peering forward from his navigator's cupola as his pilot, Flying Officer Bill Herbert from Elwood, Victoria, bore in on a merchant ship. Jack remembers that there was another Beaufighter in front of them and that 'Bill's rockets went over the plane in front and hit the ship'. The six New Zealand torpedo Beaufighters followed the Australians in and seriously damaged two merchant ships. The convoy seemed to have been caught by surprise, and it wasn't until late in the attack that the flak defences became effective. Flak from the convoy was meagre but flak from the shore built up, coming from the island of Wangerooge and from Schillighorn. Flying Officer George Hammond came on the VHF radio and warned, 'This is heavy flak. Keep weaving!' Jack Tucker remembers that they took the advice.[27]

It was another furious exchange between ships and Beaufighters. Five Beaufighters were damaged, four of them from 455 Squadron. After the aircraft broke off, they climbed to 2000 feet and set course for base; some by themselves, others in small groups. An hour and a half later they closed on Langham. Control at Langham brought some of the damaged Beaufighters down first: Proctor at 10.05 pm, with flak damage; five minutes later was Hakewell who belly-landed; 10.20 pm, Walker, escorted by Pilcher, was down, damaged but safe. Herbert and Steer came in undamaged, then at

10.35 pm, the young American Rouse landed, followed by Higgins. Cock's Beaufighter was damaged, and he came in at 11.05 pm. The last of the Australians in was their leader, Wiggins. Crews grouped in the Operations Room for a debriefing by the Station Intelligence Officer, and a nip of rum.

9 September 1944 was another big day for Langham. Fourteen of 455's crews were on operations and three Squadron aircraft were used twice. They started before first light with a reconnaissance to the Dutch coast by a single aircraft, then came a reconnaissance in force, and later they flew five separate reconnaissance patrols throughout the rest of the day. The second operation, a reconnaissance by sixteen Beaufighters, flew to the Dutch coast at four hundred feet, in close formation, with wing and tail lights on. They ran into,

a front of heavy rain, hail and almost continuous lightning which lit up the cockpit like Dante's Inferno. The lights of all the other aircraft were blotted out and fears of a collision plagued our thoughts. The leader called up and said to return to base independently. How there wasn't an accident was sheer luck. I thought of all the aircraft turning blind and decided to go down gently straight ahead. It was now gradually becoming paler, signifying dawn's arrival but was still a 'flying on instruments' task. It continued to rain heavily with intermittent lightning. I levelled out at twenty feet on the radio compass and made a gradual turn onto a reciprocal course for home. Soon the white-capped sea could be seen splashing as it whipped under the nose and then a sigh of relief the storm was left behind.[28]

The last patrol for the day was sent out just before 5 pm, flown by Flying Officer Bill Stanley of Northbridge, New South Wales, with navigator Flying Officer Ken Dempsey from North Brighton, Victoria. Their Beaufighter was hit by flak and lost an engine. Flying control was on standby as they approached Langham with the commander of 455, Davenport, waiting at the tower. Flight Sergeant Jack Tucker saw the landing,

[Bill Stanley] chose to approach on one engine correcting with his rudder. When he throttled back for touch-down the rudder correction didn't work and the plane ground-looped. It burst into flames. Ken managed to get out but Bill couldn't because his flying boots were jammed under the instrument panel. Jack Davenport tore out in his jeep, ran into the flames and pulled Bill out of his boots and the cockpit.[29]

Stanley was severely burned, and Jack Davenport himself had burns to his hair, face and hands. In the ambulance, although in great pain, Stanley reported the details of the convoy they'd found on their patrol. Davenport carried on his duties and was back flying in several days. He was later awarded the George Medal for his bravery in saving Stanley's life. Another Squadron pilot later said,

Those petrol-fed fires travel fast and the intensity of the heat has to be experienced to be believed. Anyone who saw the extent of Bill's scars later would know the crash truck would have arrived in time to remove nothing more than a well cooked crisp. Makes you proud of your CO.[30]

455 Squadron kept up the pace. Next day there were morning and evening patrols by the Langham Wing, involving nineteen crews and fourteen aircraft. Two crews and seven of the aircraft were used on both patrols, giving them a total of seven hours operational flying on one day. In the morning they patrolled for five hours off the Frisian Islands but found nothing. On the evening patrol, eight Australian crews joined six from the New Zealand Squadron and another twenty-four Beaufighters from North Coates for a reconnaissance in force to the Frisian Islands. The leader, Wing Commander Bill Tacon, Commanding Officer 236 Squadron RAF, took the formation through the gap between Texel and Vlieland islands. Jack Cox was on that operation. He says,

> Both these islands were heavily defended with Ack-Ack guns of various calibres. It was calculated that the heaviest guns on both islands when fully depressed would fall short of each other enough to allow an aircraft to be flown through without being hit if it kept very low to the water and in single file. This proved true, if somewhat nerve-wracking, to the first few crews.[31]

There was no convoy so the leader ordered an attack on gun positions firing light flak from the southern tip of Vlieland. The Beaufighters scored hits which caused an explosion followed by a large cloud of black smoke.

They were back to the southern coast of Norway the next day. Thirty-nine Beaufighters from Langham and North Coates had an intense exchange with four minesweepers. The Beaufighters started patrol off Kristiansand and followed the southern Norwegian coast, flying westward at 2000 feet. Visibility was good and they saw the minesweepers from 15 kilometres away, just short of Mandal on the very southern tip of Norway. The minesweepers were also heading west. The ships increased speed then turned to face the attack. The Beaufighters dived straight in and the minesweepers were hit with cannon, torpedoes and rockets. The second minesweeper received most attention from the Australians, taking an estimated thirty-two rocket hits, above and below the waterline. Two minesweepers were sunk and the other two seriously damaged, but five Beaufighters were damaged by flak. All the aircraft made it back to base.

It was on that evening that an American Army patrol from the 85th Reconnaissance Squadron, 5th Armoured Division crossed the German frontier near Saltzburg, five years after the Polish campaign. The War had reached German soil.

6

'Into German Harbours'

...the anti-shipping squadrons had to go into German harbours to look for them [the German convoys]. This is a significant development in anti-shipping tactics.

Coastal Command Review, September 1944

Making several dummy-runs on an escort vessel, despite severe opposition, he then attacked with the light of the moon path, scoring at least four hits with his rockets.

Citation for Flying Officer Steve Sykes after his attack on 6 October 1944.

We passed by a flak position on the far side of the island which gave us a few parting shots.

455 Squadron navigator on the wing strike against German shipping, 25 September 1944.

B Y early September 1944 the Allies were on German soil and their bombers were penetrating deep into Germany. The German garrison at Le Havre, on the eastern edge of the "Overlord" invasion area, finally surrendered on 12 September. Coastal's strike wings were causing irrecoverable damage to German coastal convoys, and the reluctance of the German convoys to expose themselves to air attack now meant that the anti-shipping squadrons had to go into German harbours to look for them. Coastal Command noted that 'This is a significant development in anti-shipping tactics.'[1]

The Dutch city of Rotterdam is the port of entry to the huge Rhine River valley and the German industrial regions of the Ruhr and Saar. Rotterdam had been held by the Germans since 1940. The Hook of Holland, a port and holiday resort, lies at the mouth of the waterway called the Nieuwe Waterweg, twenty-eight kilometres downstream from Rotterdam. Northbound German convoys would routinely leave Rotterdam and the Hook of Holland at last light and sail north to reach the defended anchorage of Den Helder before dawn. They would shelter there during the day and continue north next night. In the same way, southbound convoys timed their trips to reach Den Helder at dawn, and they would complete the last lap to the Hook of Holland and Rotterdam next night. A stretch of water known as the Marsdiep forms a two kilometre wide channel between the island of Texel and the Dutch mainland near Den Helder harbour.

On the evening of 12 September, ten 455 Squadron Beaufighters took off as part of a Beaufighter patrol to the Marsdiep. It was a large force, with thirty-nine Beaufighters from North Coates and Langham, and an escort of eight Mustang fighters. They flew down towards the Marsdiep at about five hundred feet, from the north. When they were nearly opposite the entrance to Den Helder harbour, crews saw a loose grouping of many ships. It included the hull of an incomplete destroyer, a merchant vessel escorted by an M-Class destroyer and twenty-one smaller auxiliaries. The destroyer hull had been launched from Wilton's Dry Dock Company of Rotterdam between 14 and 17 August 1944, and had remained there fitting-out until it left on 11 September, with the merchant ships. This convoy was spread between Den Helder and Texel, their bows facing east.

The wing leader for this sortie was Wing Commander Bill Tacon from the North Coates 236 Squadron RAF. Surprise was vital for success, so as they approached the Marsdeip Tacon ordered an immediate wheel to port, and an attack. Pilots had to turn tightly onto the convoy and some were left out of position. Only three from 455 Squadron were able to join in – Whishaw, Farr and Sykes. Flying Officer David Whishaw had faced heavy flak defences before, but he says that these earlier experiences,

> paled by the extent and sudden ferocity of the barrage thrown at us by ships and shore batteries as we entered the harbour of Den Helder. Coming in low over the southern end of Texel from about a thousand feet, all three of us had little time to identify and select targets, shooting briefly at shore batteries before entering the harbour. Once over water again, there was just time for rocket and good cannon attacks on a variety of shipping.
>
> When we had turned in to attack, my aircraft was pitched violently to port through about 45 degrees, presumably by an explosion of heavy flak under the starboard wing. As usual during an engagement, there was no thought of what the immediate consequences might be and in any case the plane when corrected, flew on normally.[2]

Whishaw attacked what seemed to him to be a tank landing craft and an escort. Photographs showed that he had actually attacked the uncompleted hull of the destroyer, which was left on fire.

Flying Officer Lloyd Farr had been awarded the Distinguished Flying Cross only a month before for his determination in four major strikes on heavily defended merchant convoys. This time he strafed an escort with cannon and

'I left the pull-out a bit too late': Flying Officer Steve Sykes [right, with his navigator, Flying Officer Lee Turner] inspecting the top of a trawler's mast that Sykes hit with the nose of his Beaufighter durng a strike on shipping near Den Helder on 12 September 1944. (Photo courtesy 455 Squadron)

took a heavy flak hit against the navigator's compartment. It wounded his navigator, Flying Officer Bert Osborn, in the leg and started a small fire. The Squadron Record Book records that Osborn extinguished the fire with 'the only liquid at his disposal.'[3]

The third 455 Squadron pilot, Flying Officer Steve Sykes, took the attack down very low. He said, 'I dived on a trawler, firing rockets and cannon all the way in. I left the pull-out a bit too late and collected the mast top with the nose and brought it back lying at my seat'.[4] The torpedo Beaufighters followed. One of these pilots said, 'The anti-flak Beaus did their stuff so well that, by the time we got to dropping range, we could hardly see the ships for smoke from the cannon and R.P. fire. I attacked two merchant ships together and dropped my torpedo between them but I couldn't see the result'.[5]

On his way out of the Marsdiep, Sykes joined several other Beaufighters and shot-up a shore gun position, a barracks and a hanger and buildings at the De Mok Seaplane Base. He turned to fly out to sea and his navigator, Flying Officer Lee Turner, calculated the course for base and passed it to his pilot. Sykes caught his navigator's attention and waved the aircraft's compass to him, showing him that it had been shot off its mountings. Sykes then turned to follow the aircraft flown by Lloyd Farr, who had set course for base.

The auxiliaries in the Marsdeip were seen to be heavily raked by cannon fire and some were hit by rockets. Cannon and rocket strikes were seen at the seaplane base. Balloons had been flying over six of the auxiliaries and from the destroyer hull, and one balloon fell in flames over the convoy. Farr, Sykes and Whishaw flew their damaged aircraft back to Langham. Farr and Sykes both made successful belly landings. The mast section taken from Sykes's Beaufighter was kept and is now on display in the main aircraft hall at the Australian War Memorial in Canberra. David Whishaw was able to land normally but he found a 'section of engine cowling and a couple of inspection plates missing.'[6]

For this attack, Sykes received the Immediate Award of the Distinguished Flying Cross. David Whishaw says of Steve Sykes, 'He was quite the most outstanding air combatant of my whole experience; a skilled pilot and re-markable shot who forced home all his attacks with the utmost determination and effect. If he had been in Fighter Command he would have been a fearsome opponent in air to air fighting. Although well liked and respected, he was always a quiet and thoughtful man who kept largely to himself.'[7]

Two Beaufighters from 236 Squadron were shot down. The strike leader's Beaufighter (Wing Commander Tacon) had been hit three times before it exploded. In his 'The Strike Wings', R.C. Nesbit includes an account of Tacon's escape,

> ...he floated through the air, pulling his ripcord just in time. He landed on the island of Texel, so badly burned around the eyes that he could hardly see. There he was found by German soldiers, who bundled him roughly aboard a boat which took him to Den Helder. On arrival, he was surrounded by a group of sailors who had evidently suffered from the Beaufighter attack. He was punched and kicked before more soldiers intervened and marched him off to the local goal. After medical treatment, he was taken to Dulag Luft, near Koblenz, and then to Stalag Luft 1 near Barth on the Baltic coast. He was eventually released by the Russians and quietly made his way back to North Coates. In his absence, he had been awarded the DSO.[8]

Three weeks after the fall of Paris the German Army was recovering its balance. The Germans were holding a coherent line, and they had denied Eisenhower the use of Antwerp and the Channel ports. Montgomery was pressing for a single powerful thrust to the Ruhr and beyond, but Eisenhower reaffirmed his intention of continuing the advance to the Rhine on a broad front. All through August, Montgomery had been seeking an opportunity to use his large Allied Airborne Army, either to cut off the German retreat or to clear the way for exploitation by Allied Armies. On September 10, Montgomery proposed to use the airborne forces for a bold and unorthodox stroke; an airborne landing to capture a bridge at Arnhem to be called Operation "Market Garden".

In mid-September 1944, Winston Churchill was in Quebec with President Roosevelt. They had before them a proposal by the US Secretary of the Treasury, Henry Morgenthau, that once Germany was defeated the industries of both the Ruhr and the Saar regions should be completely dismantled. Churchill put it to the British War Cabinet that the Russians would get the bulk of the machinery from these two German industrial regions to repair their own plants. Churchill noted that the consequences of the Morgenthau plan would be to emphasise the pastoral character of German life. The Morgenthau Plan was agreed to by Churchill and Roosevelt on 15 September 1944, when both men signed a programme for eliminating the war-making industries in the Ruhr and in the Saar, looking forward to converting Germany into a country primarily agricultural and pastoral in character.

The American press published details of the plan in late September. That suited Goebbels. The news of the Morgenthau plan appeared to offer proof that Goebbels was right when he declared that the Allies intended the extermination of a considerable proportion of the German people and the enslavement of the rest. Goebbels claimed that the Allies would turn the Reich into 'a potato patch' and that 'Germany has no illusions about what is in store for her people if they do not fight with all available means against an outcome that would make such plans possible.' Goebbels said this would redouble German resistance. That Morgenthau was a Jew was not missed. Berlin radio proclaimed, 'The Jew Morgenthau sings the same tune as the Jews on the Kremlin'. By coupling the Morgenthau plan with the demand for 'unconditional surrender',' Goebbels convinced the mass of German people that their only hope of saving the Fatherland and themselves lay in giving unconditional obedience to the Fuhrer and unconditional resistance to their enemies.

The Coastal Command combined strike wings kept growing in size as their targets were becoming more and more difficult to attack. On occasions the aircraft took on land targets rather than shipping. Early on 23 September, a Mosquito from the Photographic Reconnaissance Unit had been over Den Helder attempting to photograph shipping there from 2000 feet. It found cloud over the harbour and the results were not good. Next, a Mustang fighter fitted with oblique angle cameras was sent out. It made four runs through the entrance to the harbour, and, in the face of intense flak, took excellent pictures. Still another reconnaissance patrol was sent. Flying Officer David Whishaw with his second regular navigator, Flight Sergeant Bert Iggulden, left Langham early that afternoon with orders to stand off Den Helder, out of range of the light flak, and to observe the shipping inside. Whishaw says,

> Shipping was clearly visible inside the harbour and we flew up and down the coast about four nautical miles clear, hoping all German fighter aircraft were committed to the invasion areas south. The first run was completed without incident as Bert took photographs and we began noting the number and description of the ships. After turning to fly north again some tracer fire was visible from the shore batteries, arcing into the sea a long way short of the aircraft and giving us a false sense of security. This was totally shattered on the third run as a barrage of heavy flak burst all too near our aircraft, the small dirty black clouds of the explosions hanging in the air and covering us closely as the violently evading Beaufighter jinked up and down through hundreds of feet.[10]

Bert Iggulden remembers watching Den Helder Harbour from his navigator's cupola. He says, 'I was making two sketches, one [ship] contained a strange hull with one funnel, which looked phony, even slightly out of scale'.[11] Iggulden was on his second sketch when he saw some flashes and thought that they had the range. He says, 'I saw some dark red flashes and shouted 'Go' and Dave altered course about 90 degrees'.[12] Dave Whishaw continues,

> On the fourth and last run we stood out a little further, observations and photography completed and set course for Langham, whilst Bert gave details of the sightings by radio to base. Our sightings had complemented photographs previously taken from high level by a Mustang fitted with oblique cameras. Not long after landing and debriefing we were recalled to the operations room to give further information to a combined force.[13]

'A force of seventy-two Beaufighters': Beaufighter UB-P, flown by Flying Officer Austin Hakewell with his navigator Flight Sergeant Fred Sides, flying low and in formation on the way to an antishipping strike at Den Helder Harbour on 23 September 1944. (Photograph courtesy N. Smith)

That afternoon, three Beaufighter wings combined to return to Den Helder. Langham, Strubby and North Coates made a force of seventy-two Beaufighters, with eight Mustang fighters from RAF Coltishall as escort. Crews from the three wings gathered at Langham for briefing and it took nearly forty-five minutes to get all the aircraft into the air and formed up. The leader, Wing Commander Cartridge of 254 Squadron RAF, set course at 4.42 pm and led off at 200 knots towards Den Helder, some 20 knots faster than normal Beaufighter cruising speed. The large formation started to straggle.

Cartridge led them to the Frisian Islands, between the islands of Vlieland and Terschelling. The flak defence started but the strike force took no hits. At this stage, Mustang fighters from Coltishall were to rendezvous and dive bomb the shore based flak positions over Den Helder with 500 pound bombs before the Beaufighters went in. The Mustangs were not ready, so Cartridge took his huge Beaufighter formation on a figure eight pattern over the Zuider Zee.

The Mustangs bombed but by the time they had finished, the large

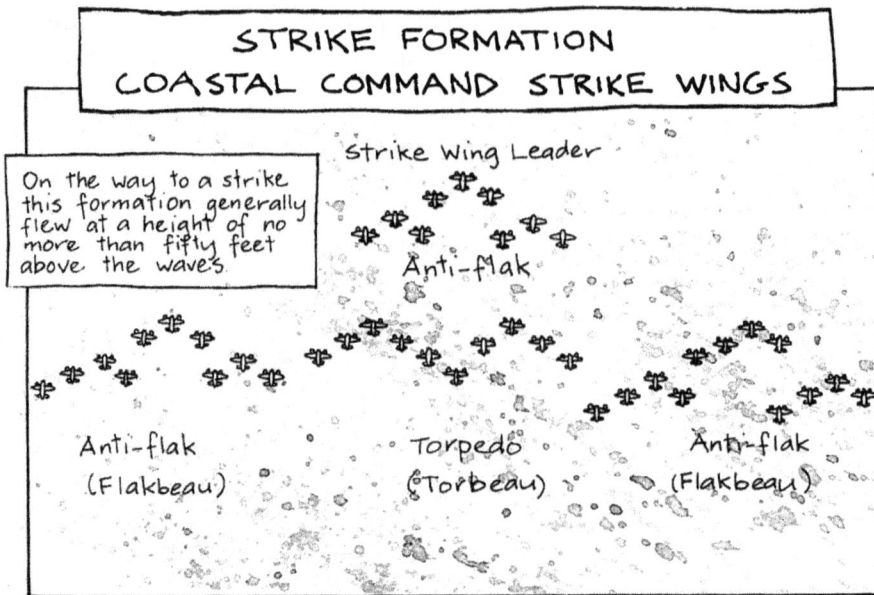

STRIKE FORMATION
COASTAL COMMAND STRIKE WINGS

On the way to a strike this formation generally flew at a height of no more than fifty feet above the waves.

Strike Wing Leader

Anti-flak

Anti-flak
(Flakbeau)

Torpedo
(Torbeau)

Anti-flak
(Flakbeau)

Beaufighter Strike Wings used this formation on the way to attack shipping. During 455 Squadron's time at Langham, a 455 Squadron pilot often lead from the front anti-flak section. Crews from 489 Squadron RNZAF often provided one of the anti-flak sections and a torpedo section. In later strikes, the Langham Wing combined with aircraft from the nearby North Coates Beaufighter squadrons.

Beaufighter force had straggled further and was not set for a properly coordinated attack against shipping. So Cartridge took them south-west over the Den Helder peninsula and towards the open sea. Some aircraft strafed land targets as they flew low across the stretch of land between the Zuider Zee and the sea. Flying Officer Neil Smith recalls that he had been low enough to wave to Dutch people at their houses, and that he hadn't felt inclined to attack anything on the land for fear of harming locals.[14] Some crews attacked a radar station, wireless masts, gun positions and a gasometer. The Beaufighters then made for base and they all landed safely. Waiting back at Langham, David Whishaw has commented that 'it was an enormous disappointment to all concerned, not the least for Bert and me, waiting for news at Langham, that the squadrons became dispersed over the Zuider Zee'.[15] Bert Iggulden has said that 'It would need to be a very unusual trip for me to remember more than a few weeks. Helder was this'.[16]

Two days later, on 25 September, another strike in the Den Helder harbour was ordered. The Langham Station Commander, Arthur Clouston writes,

> I remember the day the signal came through from Command ordering the attack on that Den Helder shipping. If I could have been participating personally in the attack, it would not have been so bad; but to have to stay on the ground myself and send off other pilots on undeniably risky ventures was one of the things I least liked about my promotion as Station Commander.[17]

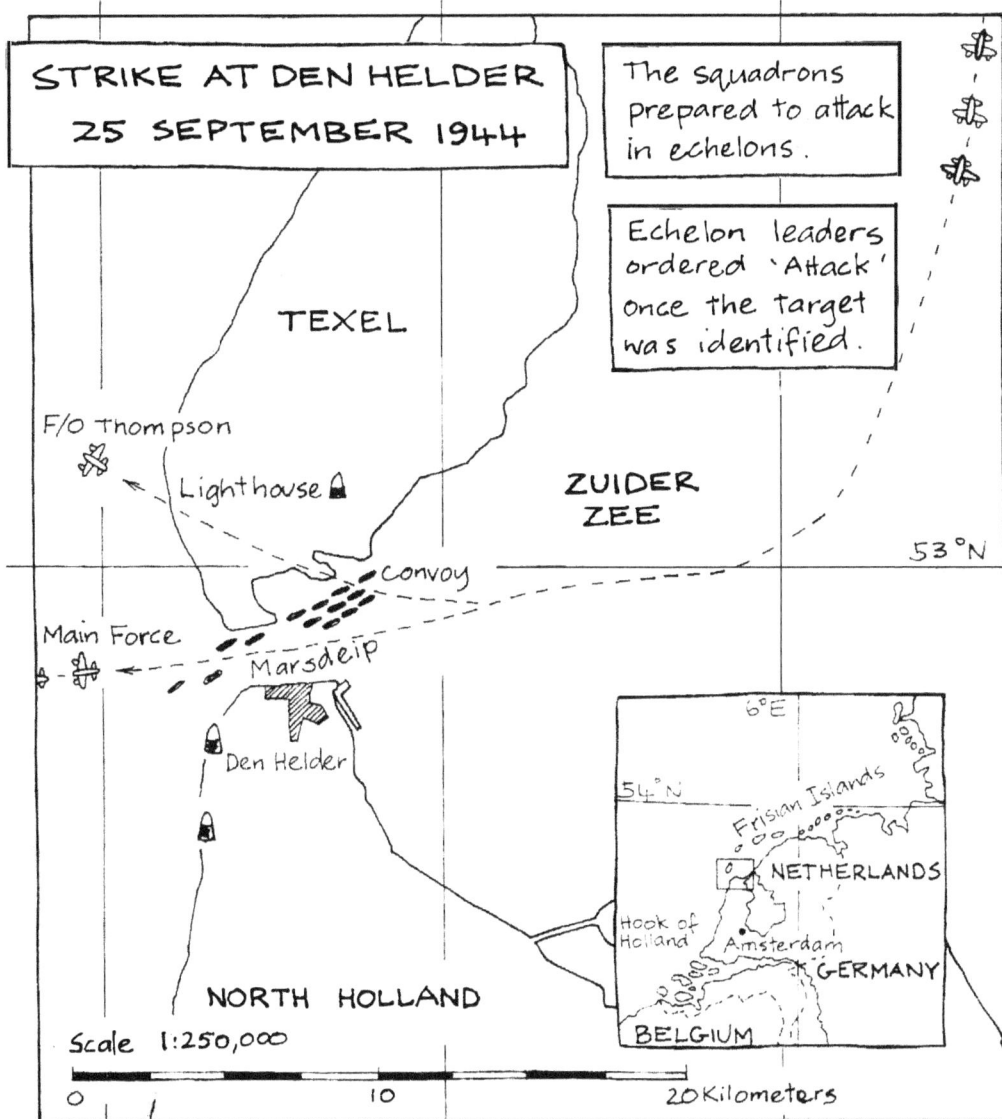

STRIKE AT DEN HELDER
25 SEPTEMBER 1944

The squadrons prepared to attack in echelons.

Echelon leaders ordered 'Attack' once the target was identified.

TEXEL

F/O Thompson

Lighthouse

ZUIDER ZEE

53°N

Convoy

Main Force

Marsdeip

Den Helder

NORTH HOLLAND

Scale 1:250,000

0   10   20 Kilometers

6°E

54°N

Frisian Islands

NETHERLANDS

Hook of Holland

Amsterdam

GERMANY

BELGIUM

16 Group organized this strike on the morning of 25 September 1944, bringing the North Coates Wing down to Langham. There was a total of sixty-five aircraft involved, twelve from 455 Squadron. Clouston led the planning at Langham and he later wrote,

> We knew that the shipping was being very zealously guarded... Obviously we wanted to catch the defence ships unawares, but eighty aircraft were by no means an easy formation to lead in a surprise attack. Even if, flying in low from the sea, the formation did take the Germans by surprise, it could only be a matter of moments before they would be subjected to the heavily concentrated fire-power of the defences. We thrashed out the possible methods of attack. The Germans would naturally expect any raid on the shipping to come from the North Sea; and that gave us our clue. We decided to attack from the rear, from the land side of the continent.[18]

Warrant Officer Ivor Gordon, navigating for Flying Officer Clive Thompson, recalls,

> We knew from newspaper reports that the operation at Arnhem by the British 1st Airborne Division was underway and it seemed that we might be able to relieve some of the pressure there. We much preferred strikes in the open sea. There was no flak from shore batteries and the Squadron could attack on a wider front from any direction. On top of that, we didn't have to worry about flying clear of cliffs or other obstacles.[19]

The briefing for the strike was held at Langham. Wing Commander Jack Davenport selected some experienced crews - his flight commander Milson, Cock (who had been with the Squadron for over a year), Payne (already awarded the Distinguished Flying Medal), Thompson and Neil Smith (now six months with 455 Squadron). Crews were apprehensive about the afternoon and Smith recalls that, 'We reckoned that it was the diciest trip that we would have to do, because of the location of the convoy and the flak.'[20] A flight commander from the New Zealand 489 Squadron, Squadron Leader Derek Hammond was to lead the strike. Hammond has been described as 'a no-nonsense man at briefings and when the target of each force had been outlined he summed up the general feeling with these succinct words: "Well chaps, there's no doubt about there being an attack. We'll go in rip, shit or bust!"'[21] Hammond was explicit about the danger of heavy flak from ships and shore batteries.

The Beaufighters left England and dropped down to about thirty feet to delay detection by enemy coastal radar. The escorting fighters seemed uncomfortable travelling at the Beaufighter cruising speed of 180 knots. It looked too slow for them and they zig-zagged back and forth to keep station.

The trip to the Dutch coast took about an hour. There was full cloud cover, and the sea was calm with little wind. They approached the Frisian Islands and the Zuider Zee from the west and flew between the islands of Vlieland and Terschelling (the gap between the islands was about eight kilometres and shore batteries from both sides opened up as the Beaufighters flew through at just fifty feet and some two hundred knots). Sykes was hit but remained with the formation. Flying Officer Jones was also hit, and with his Beaufighter damaged in the rudder, fuel tank and hydraulics, and now hard to control, he wheeled back towards base. Ivor Gordon recalls,

> As we entered the Zuider Zee a huge flock of birds rose from the water forming an enormous cloud and we flew in formation under them. We must have hit one in our slipstream. We wheeled right in an arc to the south, climbing over a causeway. I saw a cyclist on the causeway watch us fly past as Wing Commander Davenport led us up to 2000 feet. We continued the turn around to the west towards the Den Helder anchorage, this time from the direction of the mainland. It was then we saw the target; a group of...enemy ships anchored in a group near the island of Texel.[22]

There were twenty-three vessels lying at anchor in the Marsdeip, all their bows facing east. The largest was the incomplete hull of a large merchant ship,

'Shipping was very zealously guarded': This photograph was taken by 455 Squadron navigator Pilot Officer Jackson during a combined wing strike against German shipping near Den Helder Harbour on 25 September 1944. It shows the heavy flak defence from the escorts and shore batteries. (Photo courtesy S. Milson)

attended by six tugs and protected by five M-Class minesweepers and twelve auxiliaries.

Flying Officer Neil Smith was flying Q-Queenie on the port side of the formation. Q-Queenie was slower than the other Beaufighters so Smith decided that, rather than climb before the attack, he would stay at sea level and join in as they approached the convoy stretched out between Den Helder and Texel Island. It was a dull day, and crews saw many balloons above the convoy that was spread over a wide area.

Now the formation straightened out. Ahead was the hull of a six thousand ton merchant ship, fifteen escorts and an exceptionally heavy flak barrage in a cross-fire from the escorts, from Den Helder port and the island of Texel. At about 2000 metres from the convoy and 2000 feet above the sea, Wing Commander Davenport ordered over the radio 'Attack! Attack! Attack!'. The squadrons opened their throttles and increased speed into a shallow dive, pilots each taking a target to their front. At about 1000 metres they opened up with their cannons. Davenport went for the hull of the merchant ship and shot it up with cannon, then rockets, and was followed by his deputy Milson who used his cannon only. Flying Officer Tom Higgins of Warwick, Queensland, took on a minesweeper and auxiliary, and Sykes, although damaged, strafed three trawler auxiliaries. Higgins flew into a balloon cable with his port wing and the steel wire cut about a foot into the wing before it snapped. Davenport was hit in the port elevator and starboard engine. Then

'Suddenly we were low over the ships' masts': A photograph taken by Flying Officer Steve Sykes during an anti-shipping strike near Den Helder Harbour on 25 September 1944. It shows the Beaufighter flown by Flying Officer Colin Cock crashing into the sea in front of an escort ship. (Photograph courtesy N. Smith)

Cock was hit, his plane burst into flames, and it dived into the sea just in front of an escort ship. Ivor Gordon continues,

> (The cannons) were mounted in a pack of four and located under the aircraft at about midships. That put them a few feet from my feet and the din was absolutely terrific. I could see the belts of ammunition feeding down through the floor. There was no way of communicating with Tommy and all I could do was hang on and watch as an unwilling spectator.
>
> We attracted a lot of flak from the enemy ships and shore batteries which damaged a lot of our aircraft. At 300 mph our attack lasted only about ten seconds and suddenly we were low over the ships' masts. Most of 455 Squadron peeled off to the left and headed south. Before leaving Langham, Tommy and I had decided to peel off north over Texel Island, to avoid exposing ourselves to too much flak from Den Helder.
>
> Tommy banked right and we flew low over the island. He had to lift his right wing to avoid a tower on the island. We passed by a flak gun position on the far side of the island which gave us a few parting shots. We were reasonably safe by now but I remember it was one of the few times I had a real knot in my stomach. One pass at a target was the normal ration, so we set course immediately for base at Langham. We weren't required to adopt a formation on the way home, so we flew alone at 2000 feet with several other planes in sight.[23]

In the New Zealand section, Don Tunnicliffe was close behind his leader Hammond, who was planning to attack the docks and ground defences with cannon. Tunnicliffe writes,

We went screaming down from a thousand feet to hit the main port of this heavily defended area. The heavy stuff was like gigantic puff balls floating about but had to be flown through. Underneath, down to ground level, the visible tracer lanced all over the air space and it seemed a miracle would be needed to penetrate to our targets. There were so many red spitting barrels of guns all along the wharves and up on gun towers that it was only possible to squirt several targets to keep their heads down but there were innumerable others, all piercing the sky with 20mm cannon, 40mm heavier stuff and the unseen machine-gun bullets. Ships tied up to the docks were also chiming in and I emptied my cannons in one long burst along ships and docks. Then I was tearing along Den Helder's main street at second storey level and one little episode is still clear in my memory. Two girls were leaning out of a second storey window frantically waving material which could have been tea towels.[24]

Three planes were lost on this raid and seventeen damaged. During his attack, Tunnicliffe was badly hit by flak and lost consciousness. Only his Beaufighter, carefully trimmed to fly slightly nose-up, saved his life. He recovered well enough to make a miraculous landing at Langham with his left arm and left leg useless from heavy shrapnel wounds. 455 Squadron lost one Beaufighter, piloted by Flying Officer Colin Cock with navigator Warrant Officer Bob Lyneham, who were shot down by flak during their attack. Ivor Gordon says,

> Colin Cock's plane was next to ours in the Squadron formation and we must have swapped places when we peeled off to cross Texel Island. Wing Commander Davenport's navigator saw Colin's plane hit and assumed it was us. He was surprised to see us after we got back. Several days later it was announced that Colin Cock had been awarded the Distinguished Flying Cross.[25]

For his part in this attack, Gordon's pilot, Thompson, was also awarded the Distinguished Flying Cross. This had been the largest attack on enemy shipping yet made by Coastal Command. One merchant ship was left on fire after a large explosion in its hull, one minesweeper was blown up, a second left burning and three tugs and four trawlers were set on fire. Back at Langham, the surviving crews recovered. Arthur Clouston recalls,

> After a successful strike when the casualties were heavy, a beer party would take place in the mess. In fun and games and high spirits that would no doubt seem childish to the outsider, air crews found an outlet for pent-up nerves and assuaged the loss of many of their best friends. A bicycle race around the ante-room, with chairs and tables stacked in the centre, was always popular. When a dozen or more riders were getting up speed, the lights would be switched off and the cyclists would crash in a heap. We contributed regularly to the purchase of a second-hand piano. In deadly earnest, singing at the top of our voices, we would try to improve the musical qualities of the piano by methodically pulling it to pieces. Only psychiatrists would appreciate the value of such actions, and probably not even they unless they had personally undergone the experiences of those gallant and courageous crews.[26]

That night about 2,500 survivors of the British 1st Airborne Division began to withdraw south from their defensive location around the Hartenstein Hotel near Arnhem. Despite its success at Arnhem, the German Army guarding the Ruhr was at the point of exhaustion. At the same time Eisenhower lacked the resources to use this opportunity, his lines of communications having been extended in so many different directions.

The attack at Den Helder was the last operational sortie with 455 Squadron flown by Jack Davenport. Soon after he was called to visit Air Marshal Sholto Douglas, Air Officer Commanding Coastal Command. Sholto Douglas said, 'Davo, I want you to go to 18 Group in charge of operational planning for the strike wings.' Jack said that he wanted to stay with 455 Squadron. He'd served with the Squadron continuously since 1941, except for a break of three months as the Chief Flying Instructor at the torpedo school at Turnberry. Sholto Douglas told Jack Davenport that he was not asking but telling, and that Davenport was to do as he was told. Sholto Douglas told him 'You must stop shipping travelling down from Norway into the Baltic. You will have plenty to do'.[27]

The new job put Davenport at Headquarters 18 Group at Petrevie Castle, on the Firth of Forth. He was to control the operations of six Coastal Command strike squadrons, with involvement in reconnaissance squadrons, air sea rescue and fighter escort - a very demanding and time consuming job. In particular, Jack learned a lot about the Allied intelligence system. He says that the 'Intrepid' story makes clear the very great amount of intelligence information the Allies received. 'On occasion we'd get warning of a German convoy and send out a reconnaissance patrol to confirm the location. They'd sometimes come back without sighting the convoy, but we were so sure of our intelligence we would send out another reconnaissance with a strike force perhaps an hour behind.'[28] Davenport was responsible for the planning of all operations of the strike wings and so remained close to 455 Squadron. Now, Davenport left behind the responsibility of his command, responsibility he'd had for nearly ten months, since December 1943. He'd carried it out keenly, his actions setting the standards. Since taking command he had led his Squadron twenty-five times and suffered the loss of twenty-three of his men. Jack Davenport was a great leader of 455 Squadron.

On 26 September, the Allied troops trapped at Arnhem surrendered to the Germans. And that day Hitler signed a decree establishing a People's Army for the defence of German soil, by means of the conscription of every able-bodied man between the ages of sixteen and sixty.

By the end of September, Coastal Command had made five Beaufighter wing strikes on shipping in Dutch waters. Three of these were Langham and North Coates combined wing strikes of up to seventy Beaufighters. 455 Squadron has been credited with seven of the total of eleven ships sunk by Coastal Command on anti-shipping operations in September 1944. There had also been an important increase in anti-shipping activity along the Norwegian coast, with nine strikes there by Beaufighters and Mosquitoes.

September was also a critical month in the U-boat war with Germany. There was a stream of German U-boats withdrawing from their bases in the Bay of Biscay, Brest, Lorient and St Nazaire. Coastal predicted that the U-boats would spend the next few weeks redistributing, reorganising and refitting in Norwegian ports, and noted, 'When and what his next move will be is conjecture, but it is a fair bet that the Battle of the Atlantic will be resumed'.[29]

455 Squadron patrolled on alternate days for the first week in October 1944, concentrating on the shipping lanes around Borkum and Den Helder. These were daylight patrols and no shipping was found. On 5 October, Lloyd Wiggins was appointed 'Wing Commander Flying' for the Langham Wing. Lloyd says, 'the position [of Wing Commander Flying] was intended to relieve the Station Commander of some of his responsibilities regarding operations, and also to provide another leader for the Strike Wings other than those leaders already on the squadrons. In the event it became very difficult for the Wing Commander Flying to intrude on the work of the Squadrons.'[30] Squadron Leader John Pilcher replaced Wiggins as flight commander in 455 Squadron, and Colin Milson was now promoted Wing Commander and appointed the new Commanding Officer of 455 Squadron. He was twenty-five years old.

455 finally found a target during a night patrol on 6 October. They left base at 1 am and although there was some cloud, with base down to 2000 feet, moonlight gave visibility of up to six kilometres. Flying Officer Steve Sykes saw flares off the Hook of Holland, flew towards them, and found a convoy of eight to nine escorts heading north. He climbed a little to get a position fix using his GEE set (a radio navigation aid), but then lost the convoy. By then it was just after 2 am. Soon the patrol intercepted a radio sighting report from another patrol in their area, so they continued searching. They soon found a convoy of ten to fifteen ships in two columns, possibly a merchant ship surrounded by escorts. Three of the eight Beaufighters attacked.

It was Steve Sykes' first attack at night and he made several dummy runs at a minesweeper at the front of the convoy, then attacked with his eight rockets. He recorded two dry (above water) hits, and six probable underwater hits. Flying Officer Ted Watson and Flying Officer Frank Proctor, from Heidelberg, Victoria, took on the last two escorts in the convoy. Ted Watson noted in his diary,

> 6 October 1944:
> 'had to make four runs before I could get at the rear of the convoy. Flak from 15 ships pretty terrifying when going in singly.'[31]

October 15 was another busy day for the Coastal Command strike wings. With Norway becoming the focus for the anti-shipping campaigns, thirty-four aircraft from the strike wing at Banff in Scotland attacked a tanker and armed trawler off Kristiansand near the southern tip of Norway. On the same day the North Coates Beaufighter Wing took on three coasters and a light ship off Wangerooge on the northern German coast. And busiest of all the strike squadrons that day, 455 Squadron flew a reconnaissance patrol and two offensive sweeps.

'We sighted the dinghy with the two men in it': Flight Sergeant Norm Steer of Prospect, South Australia, (left) with his navigator Flying Officer Basil Roberts of Townsville, Queensland. Steer and Roberts were forced to ditch after an attack on four armed German trawlers in the Dutch shipping lanes on 15 October 1944. Although apparently safe, their bodies were later washed up on the Swedish Coast.

In the first 455 Squadron operation, at 6 am, Flying Officer Bob McColl took a reconnaissance patrol to the Dutch convoy routes. They saw some shipping in the Marsdiep but none near the Hook. At the same time, Squadron Leader John Pilcher led six crews on an offensive sweep of the Dutch shipping lanes. They found four armed coasters in line astern and four Beaufighters attacked. After the attack the Beaufighter flown by Flight Sergeant Norm Steer was seen to have a starboard engine on fire, and then to ditch. Steer and his navigator, Flying Officer Basil Roberts from Townsville, Queensland, were seen in their aircraft dinghy. They appeared unhurt and everyone thought they were safe. Allan Ibbotson was navigating for Jack Cox that morning. He says,

> Norm's plane must have been hit and from memory one engine was on fire. He ditched the A/C (aircraft) quite successfully and both aircrew got into the A/C dinghy. We circled them and waved to them. They both appeared to be uninjured so Jack (Cox) decided to complete the recce up to Heligoland. Nothing else was sighted so I suggested to Jack that I would calculate a course back to base which would pass over the place where Norm Steer had ditched. Jack agreed and sure enough we sighted the dinghy with the two men in it and apparently quite fit. I flashed 'OK' to them with the Aldis lamp.[32]

Because of the cloud base in the area was too low to drop a life-raft, the position of the dinghy was broadcast on the International Distress frequency. Allan Ibbotson continues,

Norm Steer had seen a group photograph I had with an Adelaide lass named Phyllis Long and identified her as a friend of his wife. Because both men appeared OK I wrote to Phyllis and suggested that she contact Mrs Steer and let her know that Norm appeared to be unharmed after the ditching - but it would probably be a minimum of some seven weeks before she could expect to hear about Norm through Red Cross...

We moved to Dallachy soon after that and it was while we were there that news came through to say that the bodies of Norm Steer and Basil Roberts were washed up onto the Swedish coast. I certainly regretted imparting a false hope to Mrs Steer - and am still completely puzzled as to what could have happened.[33]

After these two early operations, eleven crews and aircraft had flown, with one crew lost. In the afternoon, 455 Squadron mounted another shipping strike, with fourteen aircraft and crews. Seven crews went out for a second time that day – the pilots Watson, Smith, Cox, McColl, Herbert, McLean and Higgins. Ted Watson wrote in his diary, '1 hrs sleep then out again for maximum 'do' in Schillig Roads'.[34] With fourteen aircraft from 489 Squadron, they joined forty-two aircraft from North Coates. Lloyd Wiggins, Langham's Wing Commander Flying, led in an aircraft from 489 Squadron. Visibility was good, up to sixteen kilometres, with patchy cumulus cloud down to 3000 feet. Seas were slight. Minutes after arriving in the patrol area, they found three escorts heading due east off their starboard. The two outside anti-flak sections attacked with cannon, blowing up the E-boat and leaving a schooner-rigged sailing ship burning fiercely. A smoke screen was put into operation from the island of Heligoland and there was moderate flak from the ships and the defenders, and intensive flak from the shore. They were back at base by 8 pm.

455 Squadron had now been at Langham for six months, since mid-April 1944. They had taken part in twenty-three anti-shipping strikes, and had lost twenty-three aircrew. It was mid-October, and the Norwegian coast had again become the focal area for anti-shipping operations, with Swedish iron ore a priority target. Captain S.W. Roskill RN, author of the Official History of the Royal Navy, wrote about the Norwegian Coast,

The difficulty in extending the blockade to German ships carrying Swedish iron ore to German ports was that for almost the whole journey down the Norwegian coast it was possible for ships to remain in territorial waters by using the narrow passages between the offshore islands and the mainland, known as the Inner Leads. To us it seemed intolerable that by keeping to that 'covered way', as Mr Churchill called it, German and neutral ships could evade our control.[35]

The Russian offensive in northern Norway and the Allied successes on both the Western and Eastern fronts had also forced the Germans to start transferring troops and equipment from Norway to Germany. The RAF considered that if the strike wings could sink or halt German shipping on the Norwegian route, which now carried Germany's only import traffic as well as major trooping activity, it would seriously damage the German effort.

In mid-September, while the Langham and North Coates wings had been facing the challenge of reaching their targets in defended anchorages around Den Helder, the two Beaufighter squadrons from Strubby and the two Mosquito squadrons from Portreath switched their focus from the Bay of Biscay to the Norwegian coast. Off the southern tip of Norway, near Kristiansand and inside the fiords, the mixed Beaufighter and Mosquito forces began a new offensive. By October, five of Coastal's strike squadrons had moved to Banff, east of Inverness. In one spectacular attack, on 15 October, a formation from Banff penetrated deep into the Skagerrak and on the return journey, off Kristiansand, they attacked a laden 12,000 ton tanker and its escort. A concentration of rocket and cannon hits had the tanker engulfed in flames and heavy smoke, leaving burning oil on the sea and survivors struggling in the water. The escort, an armed trawler, exploded and was last seen blazing with decks awash. This was achieved without loss of a single aircraft.[36]

On 20 October 455 Squadron was ordered to move to the RAF Station at Dallachy, at the mouth of the Spey River some 110 kilometres east of Inverness in Morayshire, Scotland.

They left Langham in three groups, by train, by road and by air. Most Beaufighters carried a third member, a mechanic, who spent the trip sitting on cannon boxes behind the pilot. An advance party of three aircraft flew up on 22 October, two days before the main body. Flying Officer Bob McColl was with the advance party and Jack McKnight remembered him leave Langham,

> When we left Langham for Dallachy, one of our top pilots, Bob McColl, was sent up early in a three aircraft advance party. Before Bob left he did a 'beat-up' of the Langham field. I remember he came over the field really low, heading straight for me and I instinctively hit the ground as he went over. Later I was telling friends about it and said that it seemed that his propellor tips had scraped the perimeter fence. Of course the word got around and later that night I was awoken and paraded...I explained that Bob's prop hadn't actually hit the fence and that the story had been exaggerated. It caused a lot of unnecessary fuss and I was disappointed about that.[37]

The main move was made on Tuesday 24 October. As a senior ground staff non-commissioned officer it was Jack McKnight's privilege to fly up to Dallachy with his Commanding Officer, Wing Commander Colin Milson. Jack says,

> When the Squadron main body was about to depart, strict instructions were issued that there were to be absolutely no beat ups. I was to fly out in the last plane with my Commanding Officer Colin Milson. After he took off he proceeded to beat up the Officers' Mess, twice. He saw an Italian prisoner of war working in the field below tending some ducks so he beat them up too. Colin Milson flew up past the old Squadron station at Leuchars, then landed at Dallachy.[38]

After they had arrived, Jack McKnight says, 'I carefully asked him why he had done the beat up. His reply was that whatever he had done, it certainly wasn't a beat up!'[39]

7

Dallachy

Snow covered fiords with sheer sides, often plunging thousands of feet, it was hardly country for mere mortals like us.

Warrant Officer Noel Turner, describing the Norwegian coastline over which 455 Squadron operated.

The groundcrew surely earned the respect of Squadron pilots whose lives were in the hands of those dedicated people. They would await the return of aircraft from operations, anxiously searching the sky for their aircraft.

Wing Commander Jack Davenport AC, DSO, DFC and Bar, GM, MID

A T the end of September 1944, General Eisenhower had fifty-four army divisions on the Continent of Europe. But he could only deploy, on average, less than one division to each fifteen kilometres of front and he could neither concentrate enough strength for an offensive against the Ruhr, nor maintain sufficient pressure elsewhere to prevent the Germans massing additional forces in defence. Eisenhower needed the port of Antwerp to build up enough strength to reach the Rhine before the German Army could recover. The stubborn rear-guard action of the German Fifteenth Army, holding the Scheldt Estuary near Antwerp, denied the British and Americans the use of the port until the end of November, nearly three months after Antwerp city was captured. This, and the defensive victories which the Germans won at Arnhem and Antwerp would prolong the war; they would give Hitler the breathing space of the winter to fill the gap in his defences.

Now, early in November 1944, Coastal Command had two of its three strike wings in Scotland. These, both in 18 Group, were the Dallachy and Banff Wings. The Dallachy airfield was built for Coastal Command during 1942 and early 1943, the lie of the land dictating only two runways instead of the usual three. Typical of mid-war airfields, Dallachy had thirty-five diamond hard standings. Runway 228 was 1460 metres, and Runway 290

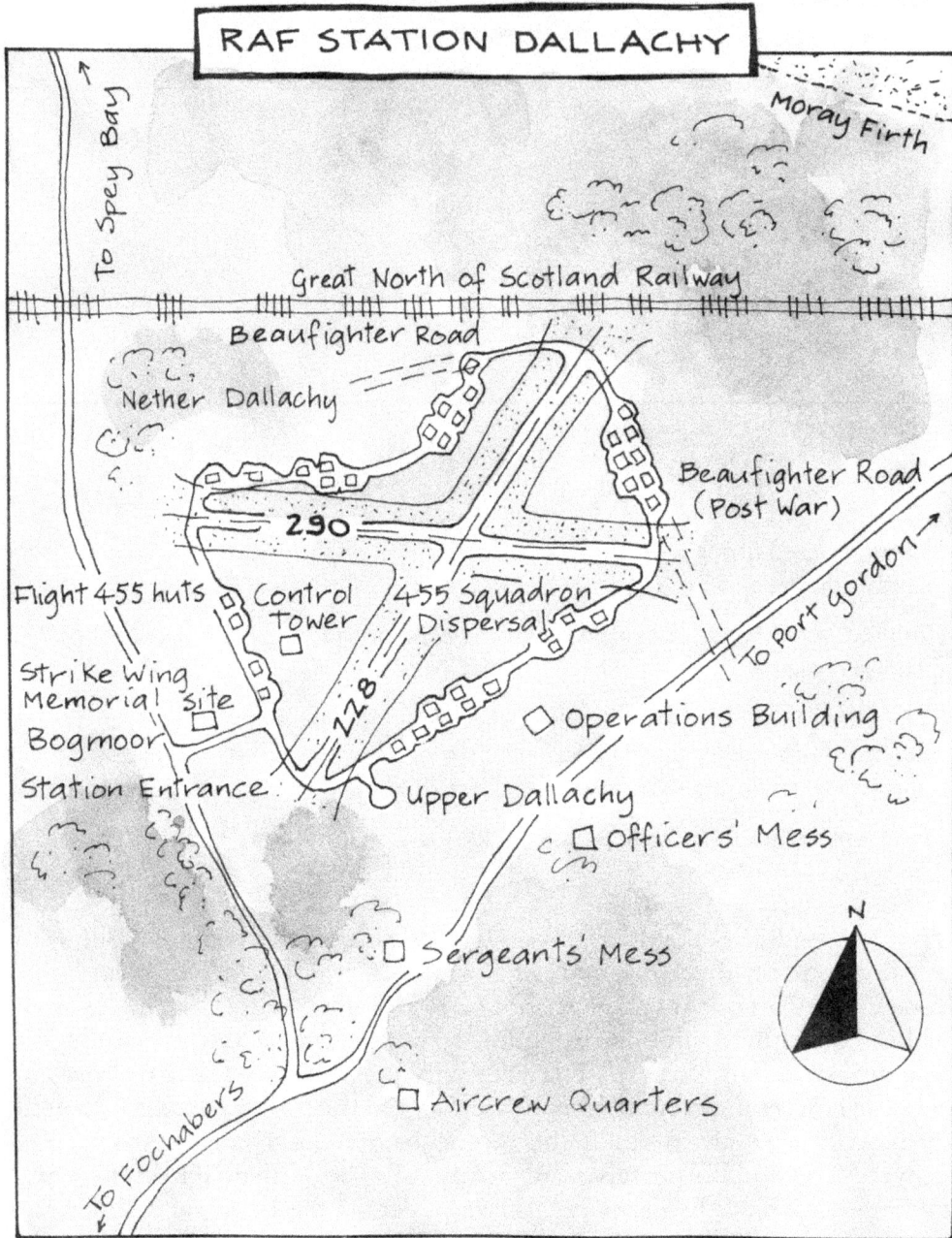

RAF STATION DALLACHY

was 1280 metres. When first constructed the airfield was loaned to Flying Training Command. Later in July 1944, Coastal Command prepared to switch some forces from the south-west of England to Scotland. Banff and Dallachy were earmarked for Wellington squadrons and Flying Training Command was informed that it must vacate these stations. It was not Wellington squadrons which arrived at Dallachy, but Beaufighters. Eventually, there were four Beaufighter squadrons at Dallachy – 144 Squadron RAF, 404 Squadron RCAF, 455 Squadron RAAF and 489 Squadron RNZAF. There was also a detachment of Wellingtons of 524 Squadron to drop flares on night operations which would assist the Beaufighters.

144 Squadron RAF had originally formed in March 1918 at Port Said, and it disbanded in February 1919. It was re-raised in January 1937, and flew in turn Ansons, Blenheims and Hampdens. 144 took part in the attacks against the *Scharnhorst* and *Gneisenau* in February 1942 and, with 455 Squadron, transferred to Coastal Command in April 1942. 144 Squadron had joined 455 on the courageous trip to Russia in September 1942, and had re-equipped with Beaufighters in January 1943. The Squadron moved around frequently for the next eighteen months, to Wick, Davidstow Moor, Strubby (to join 16 Group for the D-Day operations), Banff and finally to Dallachy in October 1944.

404 Squadron RCAF had formed in April 1941 at Thorney Island. It operated with Blenheim fighter-bombers from a number of Coastal Command stations, and re-equipped with Beaufighters in January 1943, as part of Coastal's 19 Group, providing long-range escort for anti-shipping aircraft over the Bay of Biscay. 404 Squadron was moved to Davidstow Moor in Cornwall in May 1944, then to Strubby in July for operations to support the "Overlord" invasion. They moved again to Banff for a short time in September 1944, then finally to Dallachy on 22 October 1944, their fourteenth shift of base since formation. The Squadron's crest was a Buffalo head in profile, and their motto was 'Ready to Fight'.

18 Group Headquarters was located in an underground bunker at Petrevie Castle near St Andrews, Scotland. Operational planning was an almost continuous activity there. The work involved analysis of Intelligence and reconnaissance sorties, selection of targets, co-operation with the other services, prediction of weather, consideration of enemy defences and fighters, planning air-sea rescue, discussions with station commanders and strike formation leaders, and communications. Tapes were pinned to large wall maps showing planned strikes and tension developed, particularly amongst the WAAF operators, as everyone waited for the results of attacks. Headquarters 18 Group worked closely with Coastal Command Headquarters, developing plans to deal with changing Intelligence pictures and the changing tactics of the enemy. With the breaking of German communications codes, a vast amount of information became available. At times, reconnaissance sorties were ordered even when Intelligence staff knew what the crews would find, to protect the fact that the German codes had been broken.

By now, the Germans were ordering their convoys to take shelter in the steep-sided Norwegian fiords by day, and only continue their passage along the coast in darkness. To counter, the striking forces had to be ready off the Norwegian coast before dawn to catch the shipping before it took proper shelter.

It was an early start for the Dallachy Wing on 1 November 1944. Wing Commander Colin Milson led twenty-two Beaufighters away from the airfield at 5.27 am, his first operation from Dallachy as the Australian Commanding Officer. The plan was to reconnoitre in force from Obrestad Light near the southern tip of Norway, and east coastwise around to Kristiansand. This was a vital part of the coastline. Wing Commander Jack

AREA OF 455 SQUADRON OPERATIONS
NORWAY 1944 - 1945

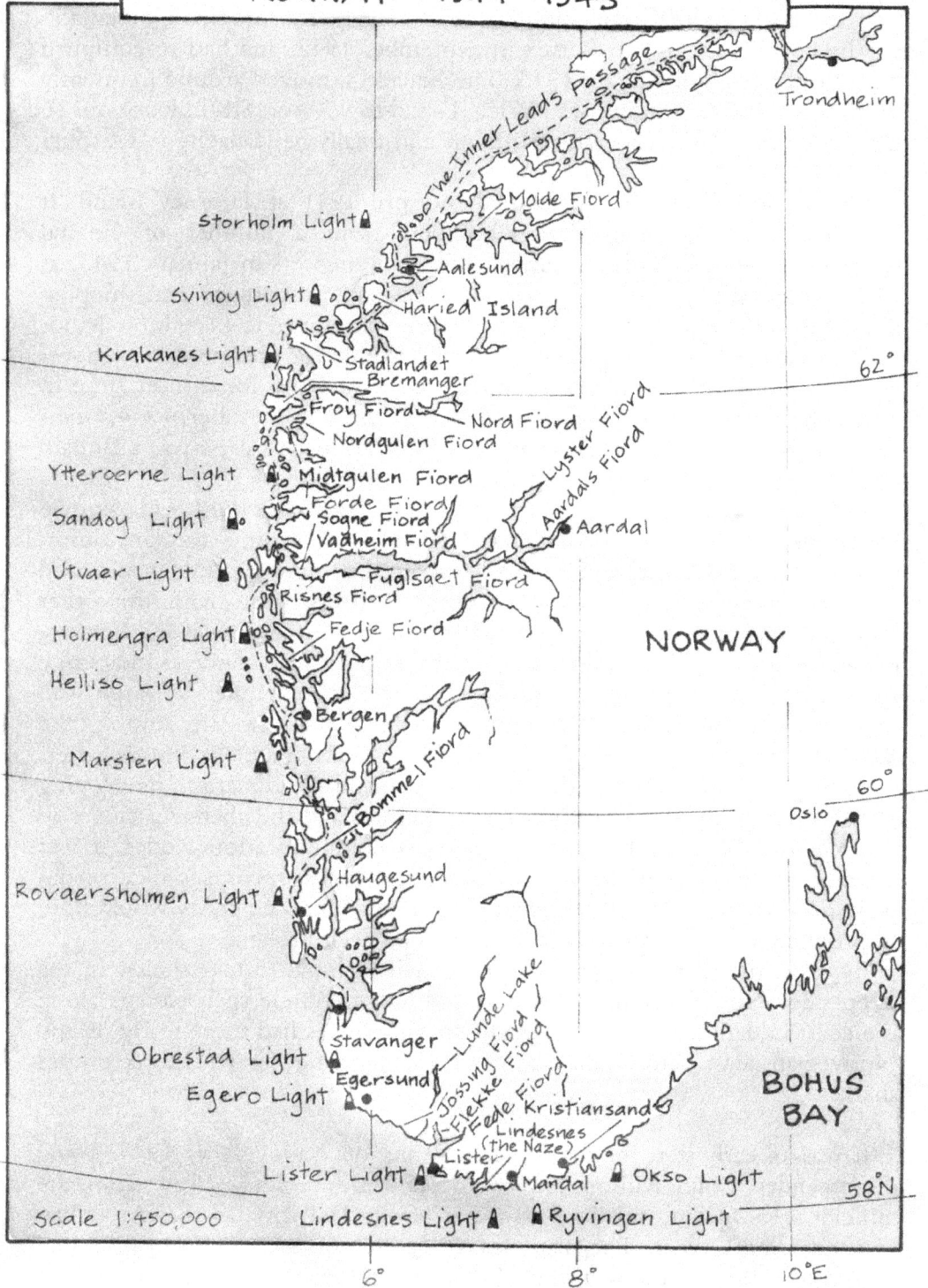

The Inner Leads Passage

Trondheim

Molde Fiord

Storholm Light

Aalesund

Svinoy Light

Haried Island

Krakanes Light

Stadlandet
Bremanger

62°

Froy Fiord

Nord Fiord

Nordgulen Fiord

Lyster Fiord

Ytteroerne Light

Midgulen Fiord

Aardals Fiord

Sandoy Light

Forde Fiord
Sogne Fiord
Vadheim Fiord

Aardal

Utvaer Light

Fuglsaet Fiord

Risnes Fiord

NORWAY

Holmengra Light

Fedje Fiord

Helliso Light

Bergen

Marsten Light

Bommel Fiord

60°

Oslo

Rovaersholmen Light

Haugesund

Stavanger

Lunde Lake

BOHUS
BAY

Obrestad Light

Jossing Fiord

Egersund

Flekke Fiord

Egero Light

Fede Fiord

Kristiansand

Lindesnes
(the Naze)

Lister

Lister Light

Mandal

Okso Light

58°N

Scale 1:450,000

Lindesnes Light

Ryvingen Light

6°   8°   10°E

Davenport, now at 18 Group Headquarters controlling strike wing operations within 18 Group, says that 'After Stavanger, the ships were forced out [of the Inner Leads] until south of Egersund. It was good to attack between Stavanger and Lister.'[1] This morning there were eleven aircraft from 455 Squadron and eleven from 489 Squadron.

The Beaufighters lined up along the taxi-ways, then turned and streamed off the ground. Milson set course over Dallachy. He had with him some experienced Australian crews, with pilots Smith, Turner and Payne. Others, such as pilot Flying Officer George Hammond, of Mosman, New South Wales, and navigator Flight Sergeant Gordon Henry of Middle Brighton, Victoria, had less than a dozen operations. Milson flew to a Drem flare some sixty-five kilometres off the Obrestad Lighthouse. The Drem flare system was named after a method of night airfield control used at Drem airfield, near Edinburgh. A circle of flame floats several kilometres in diameter would be dropped in a pre-arranged position by a navigation aircraft. Pilots would navigate to this spot and circle until it was light enough to go on patrol. An air-sea rescue Warwick was often used to lay the flares, and it would then stay in the area as safety aircraft.

It was little more than half-light, at 7.07 am on this morning when they began circling, and with navigation lights on they worked to maintain formation at 300 feet. It was a testing task, twenty-two pilots holding Beaufighters in position with throttle, rudder and control column, waiting for the signal to set course. Then disaster. Two of the aircraft suddenly converged and collided. The starboard Beaufighter from 489 Squadron did a stalling turn and dived into the sea and the other, flown by Flying Officer George Hammond, disappeared into the darkness with smoke trailing. Bright orange flashes had been seen from Hammond's aircraft, thought to be two of his rockets being jettisoned. Milson ordered an aircraft from 489 Squadron to circle the area and he called the air-sea rescue Warwick on VHF radio; the Warwick was already in the circuit but it found no trace of either Beaufighter.

Sixteen Beaufighters then left the area and set course for the patrol, from Obrestad Light towards Egersund. The weather was poor, but it later improved to give visibility of nearly 50 km when they arrived off Kristiansand thirty minutes later. At 8.13 am, at the end of their patrol area, they set course for base having sighted nothing.

There were now days spent probing Norwegian fiords for enemy shipping. Crews rarely failed to be impressed with their area of operations. One said, 'Across the water the backdrop to our activities was usually quite awesome, snow covered fiords with sheer sides, often plunging thousands of feet, it was hardly country for mere mortals like us.'[2] For another pilot, 'it was like the back of the moon'.[3]

During 455 Squadron's time at Langham, 16 Group had published a memorandum on the use of Outriders. They described a strike leader's dilemma,

> When a large striking force is patrolling the enemy convoy routes, the Leader is invariably faced with the difficulty of deciding whether to attack a small target early in the patrol, or whether to hope for a larger one further on.[4]

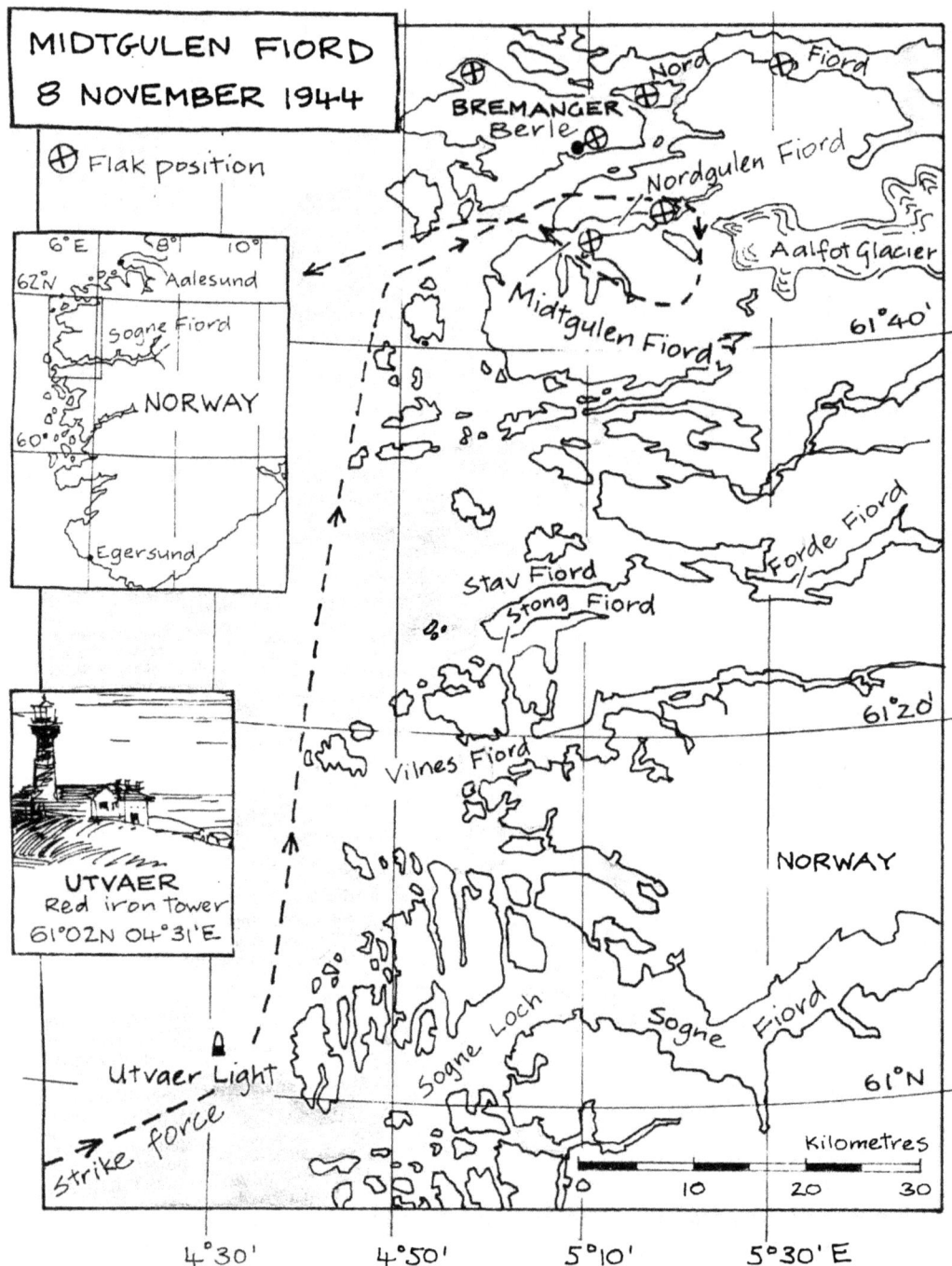

MIDTGULEN FIORD
8 NOVEMBER 1944

⊕ Flak position

UTVAER
Red iron tower
61°02N 04°31'E

The Outrider tactic, initiated by Wing Commander Jack Davenport, was to fly a single reconnaissance (later two) aircraft along the patrol route ahead of the main force to report to the leader, by VHF radio, all shipping in the patrol area before the main force arrived. The leader could then decide which target to attack. Pilots selected for this Outrider role were given special information lectures by Norwegian pilots from 333 Squadron RAF. Outrider pilots were told how to recognise important points on the Norwegian coast –

lighthouses, islands, churches, towns, railroads etc., without having to manipulate the large maps in the cockpit of a Beaufighter.[5] Jack Cox was selected to be a 455 Squadron Outrider. He says that he and his navigator would discuss silhouettes of the coast for hours with these Norwegian pilots who 'not only checked that we could recognise the coast from photographs but discussed means of escape should we be shot down, even providing us with the address of their friends in Bergen, assuring us they would provide every assistance possible.'[6]

The Outrider tactics were used very successfully from Dallachy on 8 November 1944. A patrol was ordered, with twenty-five Beaufighters and two Mosquitoes of 333 Squadron stationed at Banff travelling separately to act as Outriders. Wing Commander Tony Gadd, from 144 Squadron RAF, led the Beaufighters off from Dallachy at 9.20 am. The Australian pilots were Watson, Whishaw, Ayliffe, Cox, Higgins and Payne – an experienced section.

The Outriders made landfall near Utvaer Light, opposite the entrance to the magnificent Sogne Fiord, frequently used as a landmark for reconnaissance patrols. They flew inside the belt of islands that enclosed the Norwegian coastline – the Inner Leads – from Utvaer to Stong Fiord and Stav Fiord, but found nothing. Then, a little further north, on the northern side of Midtgulen Fiord, one of them sighted two merchant vessels. The smaller of the two was the *Helga Ferdinand* of 2566 tons. The other was the *Aquila*, of 3495 tons, belonging to the German Levante Line.

These ships were stationary, lying at right angles to the shore and some one hundred metres out. They had old fashioned stacks and carried Norwegian markings and flags. Further up the Fiord were two coaster-type vessels of about 800 tons each. The Outriders called up the Beaufighters, led them up Loch Froy, and turned starboard before reaching the heavy anti-aircraft battery stationed at Berle on the island of Bremanger. Then up onto the mountains, flying east along the north of Nordgulen Fiord and towards Aalfot Glacier. The long high ridges of the mountains above the fiord were white with snow. According to Flight Lieutenant Ted 'Doc' Watson, they flew 'over mountains, lakes and fiords about 10–15 miles inland - marvellous scenery but would not like to have to force land there.'[7] This was a new challenge; for the first time 455 were faced with shipping sheltering inside the fiords. Jack Cox says,

> We turned at right angles, to starboard, and flew south climbing steeply over the fiord cliffs near a large glacier. The Outrider informed us of the target and how to attack and escape. We turned to starboard again and flew low over the mountains behind the Outrider. We were suddenly looking down on the ships 2000 feet below. They were taken completely by surprise and we were able to attack completely unmolested. I was one of the first aircraft to attack and Allan Ibbotson made good use of his hand held camera. He took photos of the ships over the cockpit as we dived and then more over the tail as we broke away. Ack Ack shells were now bursting on the lips of the glacier (far behind us). The shells carried parachutes dangling long lengths of steel cable.

Pilot Officer 'Tiger' Payne's Beaufighter was hit, buckling his canopy and the

'Marvellous scenery but would not like to force-land there': A rocket salvo strikes the 2,566 ton Helga Ferdinand in Midtgulen Fiord during an attack by the Dallachy Strike Wing on 8 November 1944. This was 455 Squadron's first Beaufighter strike inside Norwegian Fiords.

bottom cannon plate, although it seems this was probably caused by one of the merchant vessels exploding. Flying Officer Tom Higgins' aircraft was hit by flak in the starboard tank, starboard engine and in the nose camera. Three from 144 Squadron RAF were also damaged. The *Helga Ferdinand* had been hit by several Beaufighters in quick succession and it was on fire early in the strike. The *Aquila* was raked heavily by cannon and hit by several rockets. Jack Cox concludes,

> As with most rocket attacks damage could not be assessed but a photo recce the next morning by a Mosquito revealed the outline of ships on the bed of the fiord and discolouration around them indicated they had been carrying ore. Twenty-five Beaufighter had sunk four ships without a single plane being lost. Unfortunately this strike did not prepare us for the fiord difficulties that lay ahead.[8]

For some months, aircrew at Dallachy had been following the fortunes of the German battleship *Tirpitz*. In September 1944, *Tirpitz* had been in Alten Fiord, her presence tying up Allied heavy ships and, according to Doenitz, acting as some protection against an Allied landing in northern Norway. The strike wing crews stationed in Scotland dreaded the thought of *Tirpitz* ever breaking out, knowing they may be thrown into an attack against a ship with an incredible defence against air attack. On 15 September 1944 twenty-seven Bomber Command Lancasters, from 9 and 617 Squadrons RAF, had attacked *Tirpitz*. She was hit by a 12,000 lb bomb and badly damaged, but not sunk, and after repairs was moved to Tromso. Later, on 12 November, Bomber Command attacked again, capsizing the ship. Returning from that attack, the Lancaster flown by the raid leader, Wing Commander Tait of 617 Squadron RAF, landed at Dallachy to the interest and great relief of Dallachy aircrew.

A new Squadron Engineer Officer had arrived just before the Squadron moved to Dallachy. He was Pilot Officer L.J. (Jock) Berry. Jock was a 455 original - he had sailed with the first groundcrew party when it left from Williamtown in July 1941. Jock had originally applied to join the Air Force the day after the War started, during his second year as an engineering student at university, but he was not inducted until January 1940. Then it was as an Aircraft Hand. Competition to go overseas was keen in those early days and groundcrew were selected from competitive exams. Jock was with 455 Squadron at its first station at Swinderby when it was part of Bomber Command. When 455 Squadron was transferred to Coastal Command, he was posted to 460 Squadron RAAF, remaining in Bomber Command. There, he was asked by RAAF Headquarters in London to continue his engineering study and early in 1943 he attended Heriot-Watt College in Edinburgh. In nine months he finished the Engineering degree that he'd begun in Australia. After a short commissioning course, Jock was posted back to 455 Squadron as the Engineer Officer. He recalls that he was quite apprehensive about this, returning to 455 Squadron as an officer in command of non-commissioned airmen who he had worked with and who he greatly respected. But the men of 455 were thoroughly professional and treated him with respect from the moment he arrived.

At Dallachy, each of the four Beaufighter squadrons formed their own Daily Servicing Flights under their squadron engineer officer, operating out in the squadron dispersal areas to give close support to their aircraft. Squadrons also contributed groundcrews to a Station Servicing Wing that was commanded by the Station Chief Technical Officer. Each squadron's engineer officer supported his own squadron under the direction of the Chief Technical Officer.

Before an operation, aircraft would be run up by groundcrew at the dispersal point, then handed over to the aircraft captain. Later as the Beaufighters returned, the engineer officer would categorize any damaged aircraft. This work could be difficult and wearing. Jock Berry recalls,

One frozen winter night I was Duty Engineer when the beat up squadron returned from ops. I attended all landings and categorised a/c [aircraft] not wrecked outright. Done-in, frigid, after categorising the mostly sad remainder, I reported to Wingco Mill [Chief Technical Officer] in his HQ. He shut me up and shoved me into bed, drowned me silly with Navy rum from med. cabinet, told me to write any report next day, and saw out my Duty Eng. tour for the night. Good bloke.[9]

In the early years of the Squadron groundcrew were permanently assigned to aircraft and a close relationship existed between aircrew and groundcrew. Then came 'planned maintenance'. This new scheme involved staggering the major service for Beaufighters and enabled the concentration of more groundcrew in the centralized servicing wing for major servicing. Each aircraft was scheduled in advance to avoid too many aircraft arriving for their major services at the one time. As Jock Berry explains,

> This meant that not all a/c were shoved into the sky unless a 'Maximum Effort' were ordered. All kites' flying hours were plotted on an 'Hours Chart', kites being lettered on tags to move over a rectangular graticule or network, and reflect the periodical hourly inspections to which the a/c would be submitted...All this efficiency meant that no longer was a kite's crew permanently constant; men were expected to go where existing strength permitted maximum effort. However, every effort was made to compromise, and it worked... mostly.[10]

This planned maintenance was unpopular with the Squadron because it took some groundcrew away from being dedicated to a particular aircraft. Still, Jock Berry and his crews were proud of their efforts. He notes that, 'Time and time again it was an Aussie Squadron which won the prize for serviceability without "Ops Failures".' Jock had great respect for his mechanics. One in particular was Flight Sergeant 'Cappy' Hartley Orr. Jock recalled an occasion when there was a noise in the back of a Beaufighter engine, near the super-charger impeller. Orr said he thought it sounded like a nut in the impeller chamber for the super-charger. They dismantled the engine and Orr had been right. It was, according to Jock Berry, an extraordinary diagnosis.[11]

The aircrew stood behind their mechanics. Here there was a mutual respect. One airman says, 'most of us realized that there was just one 455 Squadron, aircrew and groundcrew working together towards the one end. For our part we got a quality of servicing that was above the average and a dedication second to none.'[12] Another says,

> They [groundcrew] had to contend with all kinds of weather and around the clock activity. Their dispersal sites had very little shelter from the elements. They were in attendance at the dispersal sites for hours before an operation and for considerable time after all planes had returned from the operation. They had to contend with the constant changes with the type of armaments required, petrol loads etc., as dictated by the type of operation being mounted. It was not out of the ordinary for the groundcrews to have to arm aircraft with bombs and then be told that bombs would not be required, but rearm with rockets or vice versa.
>
> They would endeavour to have every aircraft serviceable for an operation,

especially when a maximum effort was required by Group. On one occasion when a maximum effort operation was ordered one of the aircraft burst a tyre on the way to the take off point. In order that this aircraft could take part in the operation Corporals Fisher and Badger jumped on their bicycles, rode around the perimeter track and changed the tyre in a matter of minutes.[13]

Jack Davenport pays his own tribute to the groundcrews. 'Despite the

'The maintenance achievements were quite remarkable': 455 Squadron aircrew and groundcrew at Langham. From left to right: F/O Leo Kempson, Cpl Jack Lefoe, F/O Colin Cock, Cpl H.V. Sands, F/O Wally Kimpton, F/O Frank Proctor.

incredible damage to so many aircraft, the maintenance achievements were quite remarkable... The groundcrew surely earned the respect and regard of the Squadron pilots whose lives were in the hands of those dedicated people. Frequently working in freezing conditions, strong winds, snow, rain and even hail, the serviceability records achieved were a measure of their determination, as well as their skill. They would await the return of aircraft from operations, anxiously searching the sky for 'their' aircraft.'[14]

The term 'datum line' was used by the Air Ministry in London to describe the amount of operational flying for aircrew that would constitute an Operational Tour. In 1943, after 455 Squadron had moved from Bomber Command to Coastal Command, three datum lines had been fixed for Coastal Command aircrew. For flying boat squadrons an operational tour was eight hundred hours. For twin engined squadrons on ground reconnaissance duties, it was five hundred hours and for land plane squadrons used offensively (including strike forces), an operational tour was two hundred hours. In 1944, an additional qualification was that the operational tour of duty was limited to a period of eighteen months.[15] By comparison, in Bomber Command, the first operational tour was set at thirty sorties and the second tour would not normally exceed twenty sorties.

This was a difficult issue being handled as carefully as the bureaucratic process would allow. An Air Ministry letter was sent to 18 Group on 17 November 1944, tackling the problem of aircrew failing under pressure. The Air Ministry wrote,

> In some cases, owing to the less robust constitution of an individual or his subjection to special operational strain, earlier relief will be necessary. Aircrew personnel who are relieved from operational duties on account of excessive operational strain before expiry of the normal operational period should not be sent immediately to training units where their temporary lack of zeal and zest might have a depressing effect on personnel under training.[16]

There were to be cases in the coming months where aircrew were submitted as 'tour-expired'. Usually, the crew had been in their squadron for some time, had conducted strikes against heavy opposition and had their aircraft severely damaged several times. Submissions for early relief usually contained the statement, 'There is no question of lack of moral fibre'. Groundcrew did not suffer this type of operational strain but they were still under considerable pressure. Jack McKnight recalled,

> During our busy times in Scotland the groundcrew were often overworked. We worked long hours and suffered a shortage of tools and parts but we repeatedly topped the Group statistics for aircraft serviceability. 455 Squadron often put up more aircraft than any other squadron in the Group and of the 20 aircraft in the Squadron we regularly put 16 on line and we had relatively few turn back after take-off. Often we wouldn't knock off at the end of the day but work through to get planes ready and we'd often miss meals. At one stage I worked all one day and night and then all through the next day.

I went to bed that night and woke up the next day in hospital. I was told later that I'd woken in the middle of the night and started talking about Beaufighters. An orderly took me to hospital where I had to spend a week. The Medical Officer said that I was at the end of my tether and that I needed a good long rest. He said that I could have a day off whenever I needed it.[17]

'His vigils often lasted many hours': The 455 Squadron Medical Officer, Squadron Leader Bob Macbeth, in his office at Dallachy. Macbeth always made a point of waiting in an ambulance on the runway when crews were due back from operations or training flights. Macbeth was awarded a Mention In Dispatches. (Photograph courtesy L. Macbeth)

The 455 Squadron Medical Officer was Squadron Leader Bob Macbeth who had been with the Squadron since February 1943. The men drew great confidence from the knowledge that he was always there to help them, and Macbeth made a point of waiting in an ambulance on the runway whenever crews were due back from operations or training flights. His vigils often lasted many hours. Many years later, at Bob Macbeth's funeral, a former pilot said of him,

> Bob's reassuring and quiet manner suited him to the required role of doctor but often confidant and even priest. His work across the board from flushing out worries to fishing out shrapnel earned him a Mention in Dispatches.[18]

A protective helmet known as a Flak Helmet was introduced to protect aircrew. Jack Cox recalled,

After a few crews were lost without any explanation, it was considered a possibility that pilots were being struck in the head by bullets or shrapnel and losing immediate control of their aircraft. An anti-flak helmet was devised and issued to all aircrew. It consisted of a basin shaped helmet of overlapping segments of bullet proof steel, covered with brown leather and lined with yellow suede. The pilot carried it in the cockpit and pulled it on over his flying helmet before any attack, (Some pilots claimed they placed more value on other parts of the anatomy and used the helmet accordingly).[19]

'Sykes's propellor grazed him on his Flak Helmet': Flight Sergeant (later Warrant Officer) Allan Ibbotson from Perth, Western Australia, was a navigator in 455 Squadron from June 1944 until May 1945, completing a full operational tour. Two Beaufighters collided over the sea off the coast of Norway on 25 November 1944 and Ibbotson's cupola was smashed by the propellor of Sykes's Beaufighter that passed over them when manoeuvring in close formation.

The Banff and Dallachy strike wings put up eighty-one aircraft on 25 November 1944. The Beaufighters from Dallachy flew to Banff to join the Mosquitoes and they set course together for Norway. They made the Norwegian coast at Karmoy, just south of Haugesund Harbour, and wheeled in a turn north up the coast when the weather again interfered. Flight

Lieutenant Ted Watson wrote, 'flew through most continuous rain ever. Clouds did not break until we were right on the coast and it was impossible to form up in time to attack.'[20].

Five kilometres from Haugesund they encountered heavy flak from shore batteries. No shipping was seen but the cloud, down to 1500 feet, prevented a good view. The leader radioed that the target they had been looking for, a troopship, had moved and he ordered 'Abandon! Abandon!'. Flying Officer Steve Sykes started to bear away to starboard. Slightly above him in formation was Flying Officer Jack Cox. Cox's navigator, Warrant Officer Allan Ibbotson, saw Sykes coming up towards them and he called out to Cox 'pull up!'. But Sykes' plane came up and over Cox and the propellors smashed Allan Ibbotson's cupola. Ibbotson was ducking, and Sykes's propellor grazed him on his Flak Helmet.[21] Jack Cox says, 'Warrant Officer Ibbotson owes his life to this helmet...The propellor of UB-E sliced a piece of the leather covering from his helmet as he ducked and he saw his machine gun knocked from its mounting.'[22] Sykes damaged his propellor blades and Ibbotson recovered to plot a course home for his pilot and for Sykes. In the rain and heavy cloud, the force flew back to Dallachy.

That trip was the last for Ted Watson. An original from the day the Squadron arrived at Langham, he was now to return to Australia and then, perhaps, to the islands in the South-West Pacific. He wrote next day in his diary,

> 26 November 1944:
> Col [Milson] called me into his office and told me that I was to go home as R.P. [Rocket Projectile] officer - what a wonderful break. Dave [Whishaw] said I had been recommended for a D.F.C. Certainly my lucky day.[23]

The strike wings were now hitting hard, causing considerable damage to the Germans. At the Fuhrer's conference on 1 December 1944, Doenitz told Hitler that, because of 'constant heavy attacks...off the Norwegian coast, the time is not far off when ship movements in this region will come to a complete standstill'.[24]

Nordgulen Fiord, a sheltered anchorage just south of Bremanger, well protected by steep cliffs and land-based flak batteries, was often used by merchant ships as a shelter during the day whilst travelling the Leads. A huge force of thirty-four Mosquitoes from the Banff Wing attacked shipping there on 5 December. The Mosquitoes found the ships making excellent use of the terrain, protecting themselves with the steep cliffs at the very eastern end of the Fiord. The Mosquitoes, forced to attack in steep dives and into flak from the escorts and shore batteries, still managed to leave several of the vessels burning fiercely. Soon after this strike, Nordgulen was also visited by the Dallachy Wing.

Flying Officer Austin Hakewell, known as 'Aussie' in 455 Squadron, was from Mildura in Victoria and before the war was a teacher in a farming district near Swan Hill. In the May school holidays in 1940 he enlisted in the AIF. By December in that year, Hakewell was on board the *Queen Mary* heading for the Middle East and he later served in Libya, Egypt and Syria as a driver in the 1st Ordnance Field Park. He was back in Australia in March 1942 and transferred to the RAAF. With the Empire Air Training Scheme, Hakewell finished his pilot training in Canada, at Prince Edward Island. He joined 455 Squadron at Langham in August 1944.

Hakewell's navigator was Flight Sergeant Fred Sides. Sides was from Brighton in Melbourne. He joined the Commonwealth Bank after finishing school and was in the Air Force Reserve until he was finally called up for air crew training. Sides left Australia in 1943 and went directly to Britain.

Just after midday on 5 December, Squadron Leader John Pilcher led Hakewell and sixteen other Beaufighter pilots away from Dallachy and set course for the Norwegian coast. They made landfall ninety minutes later, three kilometres south of Ytteroene Light, and there they split up. Flight Sergeant Ray Dunn was an Outrider and he flew overland on a 40 degree track, heading just east of north. A second Outrider, Warrant Officer John Ayliffe, flew a 30 degrees track at 1000 feet, and Squadron Leader John Pilcher took the main force north-east towards Nord Fiord.

Pilcher continued north, and some thirteen minutes later crossed Nordgulen Fiord, the site of the Banff Wing's determined strike less than half an hour earlier. His crews saw one of the ships well alight and a column of black smoke rising over the Fiord. The German anti-aircraft gunners, still defending themselves, laid a barrage over their position. Pilcher flew on over Nordgulen to Nord Fiord, then to Orsta Fiord where he found three merchant ships. Inside the Fiord the water was smooth, almost glassy at the very eastern end of the Fiord near the town of Orsten, and conditions were clear and bright. There were two merchant vessels on the south side and the third was at the eastern end close to the town of Orsta. The Beaufighters were in two echelons. Pilcher, leading the first echelon, took with him two crews, Proctor and navigator Jones, and Addison with navigator Mason, and they attacked up the Fiord to strafe the ship anchored near Orsten. Pilcher made his attack with cannon, the tracer arcing in towards the ship, falling short into the water first then moving up and striking the superstructure with explosions. His cannon strikes stretched over the water to the foreshore as he pulled out of his attack. Pilcher banked to port as Addison followed in and they continued in a sweep over Orsta, then turned back down the Fiord to attack the other two ships.

Flying Officer Neil Smith was leading the second echelon with Herbert and Hakewell. Smith lost the chance to attack the first ship so he formed up to the east of Orsta, then led his section down the Fiord, following Pilcher, first strafing the ship near Orsten. Flying Officer Bill Herbert in Smith's echelon remembers,

> As we approached the target, low over the town and within firing range, I pressed the cannon firing button but nothing happened. Luckily the eight rockets all appeared to score hits on the ship. It was then a case of getting out without being hit. Flying as low as possible and weaving about as we turned starboard in the fiord I remember seeing the tracers from the ground positions bouncing off the sides of

STRIKE AT ORSTA FIORD
5 DECEMBER 1944

Map of the strike at Orsta Fiord, 5 December 1944:

VARTDALS FIORD

1. Pilcher attacked first up the Fiord and strafed a merchant ship anchored near the shore, then turned to attack down the Fiord.

2. F/o Hakewell was hit by flak and his plane exploded and crashed into the Fiord.

F/o Smith

F/o Herbert

ORSTA FIORD

● Orsten

S/LDR Pilcher

944

Hakewell
Smith
Herbert

835

62°10'N

3. Smith led a second attack down the Fiord then escaped over a saddle between two hills.

Aalesund
62°N
NORWAY
8°E
60°

0   5 Kilometres   10

6°E   6° 10'

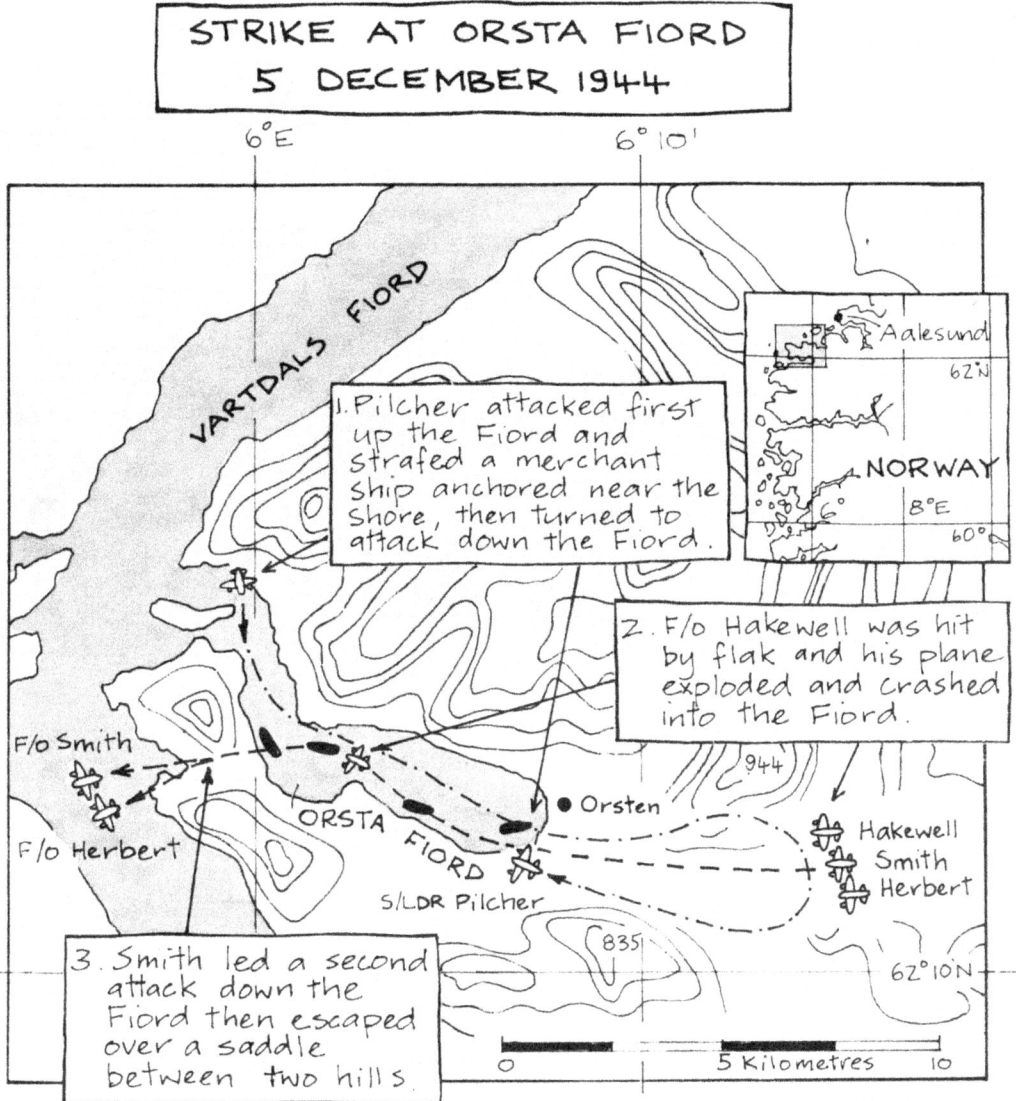

the fiord. Then quite suddenly there was a minesweeper ahead firing at the aircraft. Aussie Hakewell [in] V/455 was flying about 100 yards ahead on my port side firing at the ship when he received a direct hit and exploded in a huge orange ball of flame just above the water in front of the ship.[25]

The Outrider aircraft flown by Warrant Officer John Ayliffe arrived separately at Orsta Fiord and he joined in the attack with Smith. Ayliffe says,

On rounding a small headland I noticed another Beaufighter on my port side. As I flew alongside of this aircraft I saw a minesweeper steaming towards[us]. This minesweeper had a heavy calibre gun on its bow and was firing at the aircraft. Just as I lined up to attack I witnessed the other Beaufighter receive a direct hit and explode in mid-air. I have a vivid impression of the forward part of the fuselage and wings going in one direction and the rear portion of the fuselage going in the other direction. I am sure that the plane was flown by Aussie Hakewell.[26]

'He banked left and made for a low saddle': Flying Officer Neil Smith follows Flying Officer Bill Herbert to strafe a ship in Orsta Fiord on 5 December 1944.

Neil Smith, leading the second wave, was now flying low, close to the waters of the fiord. Sweeping on down the waterway he banked left and made for a low saddle between two hills. In front of him there was a smaller ship near the shore, under steam. Smith and the crews with him attacked with cannon, swooping across the ship and over the saddle of land.

After his attack Warrant Officer John Ayliffe made for the coast and set course for base in company with a Beaufighter in distress, flown by Pilot Officer Arthur Winter from Sydney. Ayliffe led towards the Shetland Islands and as they approached the base at Sumburgh, Winter was forced to ditch. Ayliffe saw Winter and his navigator, Pilot Officer Cliff Dunshea of Maryborough, Queensland, climb into the dinghy and after thirty seconds Winter's Beaufighter sank. Ayliffe circled the dinghy for thirty minutes until the air-sea rescue Warwick arrived.

The episode at Orsta Fiord did not end that day. A young man named Sverre Hagen was watching the strike from his house on the north side of the Fiord. He saw Hakewell's plane crash in the middle of the fiord, opposite his house, and he rowed out to the spot. He found only a parachute floating on

Beaufighter navigator Flight Sergeant Fred Sides from Brighton, Victoria, whose aircraft was shot down during a strike against shipping at Orsta Fiord on 5 December 1944. Both Sides and his pilot, Flying Officer Austin Hakewell, of Mildura, were killed.

the water, collected it and hid it from the Germans. In 1947 the parents of Fred Sides travelled to Orsta Fiord to visit the spot where their son had died. The Beaufighter was retrieved from the Fiord – there was little left but tangled wreckage. On that spot they held a memorial service and dropped flowers onto the waters. At a place on the shore, on the north side of the Fiord near where Sverre Hagen saw the crash, a memorial stone was unveiled. The inscription on the stone reads,

> *Our beloved boy Frederick Gordon Sides Australian Air Force lost in Orsta Fiord 5 December 1944. Erected by his sorrowing parents and sister on their visit from Australia June 1947. A tribute to his friend Austin Hakewell.*

During that trip to Norway in 1947, the Sides family were given the parachute that had been collected from the Fiord. Fred Sides' family revisited Norway in 1948, and his father and sister again in 1960. Since then, his sister has returned a further three times to renew the friendship with the people of the town of Orsten. Now each year on 17 May, Norway's National Day, residents of Orsten gather at the memorial site for a simple ceremony to pay tribute to the sacrifice made by Fred Sides and Austin Hakewell.

On 9 December, seven 455 Squadron aircraft patrolled to Norway in company with Beaufighters of 144 and 404 Squadrons, escorted by Mustangs

'Each year...residents of Orsten gather at the memorial': Mr Sides Senior beside the memorial erected to the memory of his son Fred, killed in Orsta Fiord on 5 December 1944. The memorial overlooks the stretch of water where Flying Officer Austin Hakewell and Flight Sergeant Fred Sides crashed in their Beaufighter after being hit by flak.

of 315 (Polish) Squadron from Fighter Command's 13 Group base at Peterhead, north of Aberdeen. As they were approaching the Norwegian coast an Outrider reported a target to the leader. Pilots strained for a view of the Norwegian coast and as it grew larger they searched for the familiar shape of Utvaer Light. Four minutes later the main force arrived at Utvaer and wheeled north, up past Sandoy Light, along the incredible coast with so many fiords and on to the distinctive Yttereorne Light. Here they turned back out to sea, climbed, flew south again and met the Outrider which led them back to his target. It was an unescorted merchant ship, the Norwegian *Havda* of 678 tons, travelling south near Vilnes Fiord. The ship had a grey hull, light superstructure and a black funnel displaying a Swastika in an orange circle.

Six 455 Squadron Beaufighters dived across the beam of the ship with rockets and cannon firing, down to 300 metres. It was almost a procession, the Beaufighters taking it in turn, flying into the sea mist towards the Norwegian coast. A crew from the Canadian 404 Squadron, on only their second operational sortie, took their Beaufighter down so close that their aircraft hit the mast. The wing was torn off, and they hit the sea and exploded in front of the target. Sykes went in, strafing across the centre of the ship, with Thompson who shot up the bow, and Whishaw who riddled the ship with cannon strikes just as the Canadian Beaufighter crashed and exploded to starboard. A pall of smoke several hundred feet high formed over the ship as the Beaufighters continued to stream in. In the end the ship was almost obliterated, and as Smith attacked he could see only the front section. Some aircrew recalled that at the briefing before the strike, they had been told that

'It deserved special treatment': Flying Officer Neil Smith fires a salvo of rockets at the merchant ship Havda *off the Norwegian coast on 9 December 1944.*

the ship may have been carrying senior Nazi officials, so that it deserved special treatment. Three Beaufighters went back for a second attack with cannon. Overall, nearly sixty hits were scored with rockets and the ship was left blazing, almost stationary and turning to land.

Flight Lieutenant David Whishaw flew Beaufighter S-Sugar for the last time that day. He says it was 'a beautiful aircraft which arrived brand new during the early days of the conversion from Hampdens...She was smooth, fast and trimmed perfectly to fly for miles 'hands off'. Other new aircraft arrived from time to time but never another like S-Sugar. Eventually, through damage and many hours usage, she faded a bit and became less reliable'. Whishaw says that this day, 'S-Sugar was feeling her age badly, but she didn't let me down.[27] This was also the last strike against shipping by David Whishaw. He was posted to the RAF College at Cranwell for a Junior Commander's course and then sent back to Australia, to RAAF Base Laverton, in preparation for a posting to Japan.

There were no more strikes against shipping by 455 Squadron in December 1944. As 1945 approached, the northern winter would continue to be a trial for the strike wings, but it would never eclipse the dangers presented by the steep-sided Norwegian fiords, the flak defences from ships and shore, and the increasing intensity of attacks from German fighter aircraft.

8

Fiords and Enemy Fighters

Don't panic! Stick together and give the escorts a chance.

18 Group Instruction to aircrew on dealing with attack by enemy fighter aircraft

I could see it [the flak] hitting my wings, coming closer to the cabin. Then as the flak hit the cabin, I fired my rockets.

Flying Officer Harold Spink, on the strike at Forde Fiord, 9 February 1945

THE German Wehrmacht's last flicker of strength was in the Ardennes forest area in Belgium, in December 1944. This offensive, their final gamble, began on the 16th. Hitler still believed that, under a few more heavy blows, 'at any moment this artificially-bolstered common front [of the Allies] may suddenly collapse with a gigantic clap of thunder.' Hitler told his audience at a conference on 28 December 1944, 'we must allow no moment to pass without showing the enemy that, whatever he does, he can never reckon on capitulation. Never! Never!' Hitler's gamble was to drive west and recapture the Allied supply port of Antwerp by cutting through the Ardennes and across the river Meuse. This would cut Eisenhower's forces in two at the weakest point on the Allied front.

The Germans had initial success, due to the resilience and courage of their Army, but by the the end of December the battle was nearly over. Germany no longer had the power to repeat its performance of 1940. Hitler had under-estimated the strength of the Allies, and the Germans were restricted by a shortage of petrol.

Admiral Doenitz, Commander-in-Chief of the German Navy, still had strong cards in his hand and he could be depended on to play them. Germany was building new U-boats, the Type XXI and Type XXIII. Over 120 of these new and faster *Schnorchel* equipped submarines had been launched by the end of 1944 and were due to become operational.

Christmas 1944, and the Australians at Dallachy made their own Christmas card using an airgraph. It said, 'A Merry Christmas from the Viking Boys'. On Christmas Day itself a strike had been notified but called off. Doc Watson wrote in his diary,

> Monday 25th:
> 'Trip off – NCOs in Officers' Mess – served dinner Airmens' Mess. Party in the Sergeants' Mess – bags of whiskey – Xmas motto, 'Grog's the shot'. Dance at night in Keith with Cappy Orr and Stan Young in Dave's car.'[1]

The weather during this Scottish winter was terrible. After 9 December 1944, there was a month when 455 Squadron made no attacks. Then at last, on 6 January 1945, they were given an anti-shipping patrol. Milson led off, with Cox, Payne and Farr, and shortly after came two newer crews, captained by Newell and Mutimer. They circled Dallachy, forming up with the Beaufighters from 144 Squadron RAF and a Mustang fighter escort from 315 (Polish) Squadron at Peterhead. It took over twenty minutes to get ready over Dallachy before Milson could set course for Norway.

They made the Norwegian coast at Helliso Light. This light was a tall red iron tower, painted with two white bands, on a low island off Fedje Fiord. Nearby, a little closer to the mainland in Fedje Fiord, they found a self-propelled oil-burning barge. The Outrider from 144 Squadron fired green flares over the barge. Three of the Squadron's most experienced crews – Payne, Cox and Farr went to attack. Payne had over twelve months in the Squadron and was already decorated, Cox had over six months on operations and Farr, who had joined the Squadron when it was flying Hampdens more than fourteen months earlier, had also been decorated. They attacked with rockets from 1000 metres and then with cannon from 600 metres. Cox was close when his rockets went, and the eight rocket trails closed in on the front half of the ship. The damaged barge was left adrift and almost aground. Milson then took the rest of the formation to patrol around the Fiord. They returned to circle Helliso Light, then back to the place where they had attacked the barge. They found a merchant ship very close inshore and circled for attack but a heavy snow-storm had broken and after three passes Milson had to call off the attempt.

Two days later, at just after midday, Flight Lieutenant Bob McColl was with five other Australian crews on a patrol to the Norwegian coast. They flew north past Rovaersholmen Light, the dark tower near Haugesund, led by Squadron Leader Christison from 404 Squadron RCAF. Then up nearby Bommel Fiord, passing many islands and the extraordinary scenery, up the Inner Leads and then around the east side of Stord Island attracting light flak from a position at Leirvik and heavy flak from Hauge. Now they were into Bjorne Fiord.

Christison led on a short circuit around Bjorne Fiord and saw a 500 ton tug towing a large barge and a two thousand ton merchant vessel. The tug towing

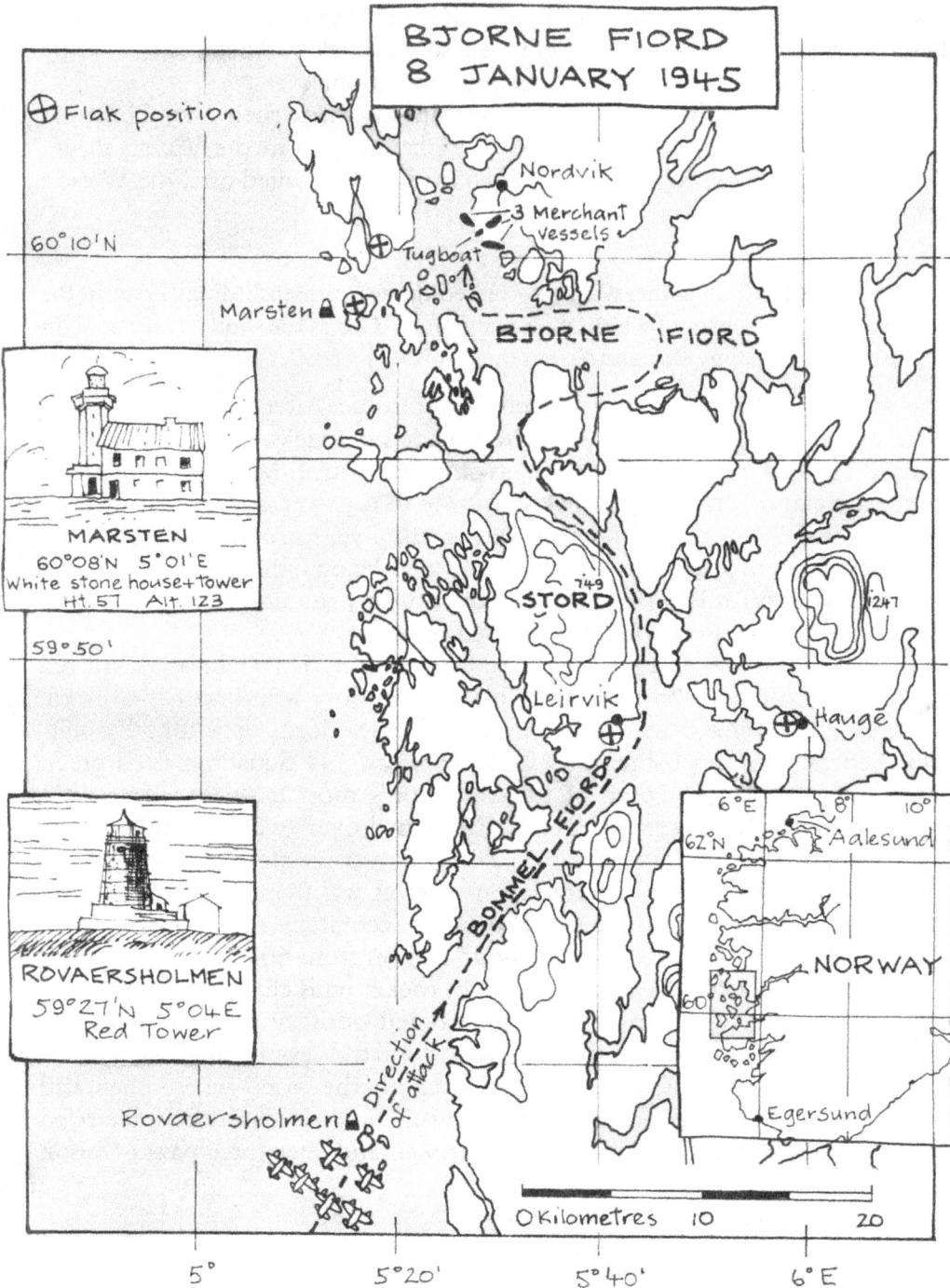

BJORNE FIORD 8 JANUARY 1945

⊕ Flak position

Nordvik

3 Merchant vessels

60°10'N

⊕ Tugboat

Marsten ▲ ⊕

BJORNE FIORD

MARSTEN
60°08'N 5°01'E
White stone house+tower
Ht. 57 Alt. 123

59°50'

STORD

749

1247

Leirvik

⊕ ⊕ Hauge

6°E 8° 10°

62°N Aalesund

BOMMEL FIORD

ROVAERSHOLMEN
59°27'N 5°04'E
Red Tower

NORWAY

60°

Direction of attack

Egersund

Rovaersholmen ▲

0 Kilometres 10 20

5° 5°20' 5°40' 6°E

the barge was well exposed, and the merchant vessel was anchored close to a rocky outcrop on a finger of land that pointed into the Fiord. These three vessels were about four hundred metres apart. Conditions were clear, with a calm sea, high cloud, and excellent visibility.

The Beaufighters raced in and light flak appeared. McColl took the target to his right, the merchant ship behind the rocky outcrops, which was making

140

'A short circuit around Bjorne Fiord': Flight Lieutenant Bob McColl attacks the merchant vessel Fusa *in Bjorne Fiord on 8 January 1945.*

dark smoke from the funnel. He strafed across this ship, his cannon shells exploding on its decks and hull, and soon only the bow could be seen for thick smoke. To McColl's starboard, the tug and barge were left enveloped in smoke. The Beaufighters wheeled to port, and a little further north again they found another merchant ship, a 1200 ton ship riding high in the water and preceded by a small trawler. They attacked again with cannon and rocket; crews saw the rockets go straight through their target. They wheeled out by Marsten Light and set course for home.

Next morning, Squadron Leader John Pilcher led to Norway again, back to Helliso Light. They flew on north-east towards Sogne Fiord in bright, clear conditions, and nearly seventy kilometres overland found a merchant ship in Fuglsaet Fiord. Fuglsaet is a small fiord some fifty kilometres from the mouth of Sogne Fiord, on its southern side. Pilcher came around from the south, flying down Fuglsaet towards Sogne. The merchant ship, the 938 ton

Beaufighters wheel away after a strike in Bjorne Fiord on 8 January 1945. The photograph shows, on the left, the ship being attacked by Flight Lieutenant McColl and on the right, a tug and barge covered in smoke. (Photograph courtesy J. Davenport)

Sirius, was anchored close to the eastern shore of the Fiord, near the town of Hoga. Heavy snow covered the steep hills around the Fiord and the town. The ship was lying amongst ice, the stern only metres from the shore. The Beaufighters went in over Hoga and as Pilcher let his rocket salvo go the name *Sirius* was clearly visible on the port side of the hull. The ship answered with some light flak but this ceased as Pilcher's eight rockets slammed into the side of the *Sirius*. The Squadron Operations Record Book has this entry,

9 Jan 45:
Results Observed: Vessel heeled over to port at angle of 45 degrees. Decks smashed and blackish brown oily smoke rising to 200' and fire broke out from aft funnel.

The *Sirius* later sank.

While the Dallachy Wing was in Fuglsaet Fiord, seventeen Mosquitoes from the Banff Wing attacked merchant ships and auxiliaries at the top of Bommel Fiord, off Leirvik on Stord Island. The Mosquitoes hit and damaged the five thousand ton *Claus Rickmers*, leaving her on fire. They returned to Leirvik six days later and attacked the *Claus Rickmers* again, but as they were breaking away they were intercepted and badly mauled by nine Focke-Wulf 190s. Five of the Mosquitoes were lost.

Flight Sergeant Peter Ilbery and his navigator, Flying Officer Bill Bawden, had both joined 455 Squadron during the last days at Langham in October 1944. Their posting to 455 Squadron was the result of an unusual encounter with Ilbery's future Commanding Officer. Ilbery was on leave, staying in a hotel in Scotland.

'Eight rockets slammed into the side of the Sirius': Rockets are fired at the 938 ton Norwegian Merchant Ship Sirius *during a strike at Fuglsaet Fiord on 9 January 1945. (Photo Courtesy W. Waldock)*

At dinner one night there was only another man and a couple in the room, the war having severely depleted patronage of highland hotels. To my surprise the waiter, unasked, brought me a drink. It was, he said, courtesy of the couple in the room. I was in uniform and assumed they were acknowledging me as a serviceman. We went out for coffee and the man wanted to know about me. In the course of conversation I explained Bill's and my frustration at having to wait to get into the air and then probably as bomber pilots after having been trained as fighter pilots.

Next morning I met the couple again. This time he was in uniform, a Wing Commander in the RAAF! His name was Jack Davenport, the Commander of 455 Squadron and he was on his honeymoon. It was perhaps no coincidence that a month later Bill and I were posted to RAF Kidlington, near Oxford, to begin flying training at 20 Advanced Flying Unit.

Three months were spent there in familiarisation with twin-engine flying on the Airspeed Oxford. Then several weeks were needed to gain a second class navigator's certificate from 3 S of G.R [School of General Reconnaissance at Squires Gate near Blackpool] before going to operational training. It was now time to acquire crew from the school's navigator pool. In the pool was Bill Bawden and having got to know him in Australia early on in training I asked him to be my navigator. Because he was in his thirties he had just scraped into aircrew and I admired his attitude for 'sticking his neck out' especially with a wife and child. I suspected from the vantage of his mature years, he considered me to be as responsible a young pilot as he might be lucky enough to find and accepted.[2]

After two months converting to Beauforts and then to Beaufighters at 132 Operational Training Unit at East Fortune, Peter Ilbery, Bill Mitchell and their navigators were posted to 455 Squadron at Langham and almost immediately they joined the Squadron in its move to Dallachy. At Dallachy they continued training with other Squadron members. Peter says,

We did no real formation or low flying until we got to 455 Squadron. Noel Turner took me out at one stage, to guide me for some practice flying. I remember coming in over the land at a very low height to find myself flying at less than roof top straight over Lossiemouth airfield [west of Dallachy], and hoping that as perpetrator of this unintentional but nevertheless forbidden 'shoot-up' that my number had not been read. It was a great thrill to fly along the cliffs and then drop over the edge towards to the sea.[3]

At Dallachy, on the morning of Wednesday 10 January 1945, Peter Ilbery was waiting on standby with the crew from Beaufighter M/455, Pilot Officer Arthur Winter and his navigator, Pilot Officer Cliff Dunshea. Winter and Dunshea had now recovered from their recent ditching off the Shetland Islands on 5 December 1944. The men were playing a traditional pub game with coins called 'shove ha'penny' when they were called to the Operations building. Eighteen 455 Squadron crews studied the large map on the wall of the operations room. Peter recalls,

The green marker tape stretched out from Dallachy along the coast of the Moray Firth, past the Mosquito base at Banff, and the fighter base at Fraserburgh, into the North Sea. But instead of the usual easterly direction it headed into a longer north east path to a landfall several minutes flying above Stadlandet, half way up the Norwegian coast.

A morning reconnaissance had found a concentration of shipping near two small islands, Lepso and Haramso, just north of the town of Aalesund and close to Storholm Light. A Beaufighter formation led by Squadron Leader Christison of 404 Squadron RCAF, took off from Dallachy at 11.50 am and set course for Banff at 12.06 pm. Peter Ilbery was in a Vic with Arthur Winter, led by Flight Lieutenant Frank Proctor, on his fortieth and last operational sortie with 455 Squadron. Peter continues,

The hour or so journey across the desolate waters of the North Sea was always made at wave-top height in order to remain below the German RDF [radar] for as long as possible in order to effect surprise in the attack, as well as attempting to avoid fighter interception. On this occasion more than usual attention was paid to keeping low down on the water, with the wave-top indicator (radio altimeter) consistently reading in the red at thirty feet! Maintaining this height above the sea and keeping formation demanded concentration and crowded out uncertainties about what was to come.[4]

On the way to Norway the weather was clear, the sea calm. They made landfall at Lepso Island, eighteen kilometres north of Aalesund, and straightaway sighted a 750 ton merchant ship, with its wheelhouse and smoke stack amidships and a gun forward. The Beaufighter crews heard *Achtung Beaufighter* on their own listening-out radio channel. Then they saw a minesweeper heading for a jetty at Haram on the southern tip of Haramso Island. The leader took the formation around and the individual Vics picked their targets. Peter Ilbery says,

In readiness my safety catches were off, the cine camera in the nose activated and yet our leader still appeared undecided. No 2 [in Proctor's Vic, Pilot Officer Winter] peeled off and indecision changed to commitment as I belatedly made up my mind to go with him. A quarter of a mile behind him it was apparent that the approach to the target was a long one. Then it became clear that the vessel chosen was the M-Class minesweeper...and the flashes from the neighbouring shore revealed the support of land batteries.

As he followed Winter's Beaufighter, Peter Ilbery saw it enter the thick of the smoke and tracer from the anti-aircraft barrage. Ilbery continues, 'Watching my own tracer streaming past him on its way to the minesweeper I saw smoke billow from one of his engines and then to my dismay flames appear. As he started to torch he disengaged to the right while his rocket trails sped on towards the vessel.'

Peter Ilbery says that, at the time, his feeling was one of fury.

Instead of aiming the rockets at the waterline I waited until there was just room to avoid colliding and fired them into the superstructure...I skidded and jinked the aircraft across a flat area occupied by barracks and made for the protection afforded by a low hill to the right.

The Beaufighter flown by Arthur Winter was seen to climb to starboard, with the port engine on fire, and over the small hill ahead. Crews saw it roll on its back and dive into the water on the north-east side of Haramso. Peter Ilbery concludes, 'Our climbing over the low hill on the other side of the island over which we flew revealed a circular patch of burning fuel'. This was less than two hundred metres from the shore, and the black smoke from the fire on the water was a sharp contrast to the grey sea and the white of the snow covering Haramso Island.

The merchant vessel that was attacked had exploded and it was finally enveloped in black smoke and left burning furiously with floating debris. The

'Beaufighter crews heard Achtung Beaufighter*': Cannon shells strike an M-Class minesweeper near a jetty on Haramso Island on the Norwegian coast near Aalesund. Twenty-six Beaufighters from Dallachy attacked German shipping sheltering amongst Norwegian islands on 10 January 1945.*

minesweeper was left smoking heavily, a pall rising to 700 feet. The Beaufighters set course for Dallachy. Low on fuel, Peter Ilbery was forced to land at Sumburgh in the Shetland Islands. Arthur Winter and Cliff Dunshea were posted as missing.[5] They had joined 455 Squadron in July 1944; it was their twenty-first operational sortie.

In October and November 1944, the British Home Fleet had conducted sweeps along the Norwegian coast between Stavanger and Lindesnes, sinking two merchant ships and four escorts. Through a heightened fear of an Allied

'Winter was seen to climb to starboard...and over the small hill': A patch of burning fuel at the spot where Pilot Officer Arthur Winter crashed into the sea after the attack at Haramso Island on 10 January 1945. Winter and his navigator, Pilot Officer Cliff Dunshea, were both killed. (Photograph courtesy J. Davenport)

descent on Norway or Denmark, and to combat the effects of the strike wings, Germany reinforced its fighter strength along the Norwegian coast. By March 1945, there would be some eighty-five single-engine fighters distributed across ten major airfields in the area south of Trondheim. Germany also brought in night-fighters to combat the night-time anti-shipping sweeps by Coastal Command. The anti-shipping crews now found that their formations were being met by large numbers of fighters, up to thirty at a time. 18 Group laid down instructions for its anti-shipping crews, in the event of an attack by enemy aircraft. *Straggling must be prevented at all costs. Aircraft attacked should not turn away from the formation but should evade by corkscrewing violently. Stay together at all costs...Don't Panic! Stick together and give the escort a chance.*[6] In January 1945, with only one Mustang squadron available for escort, the Banff Wing's Mosquitoes were sometimes tasked to switch to the fighter role.

The day after Haramso, on 11 January 1945, thirteen Mosquitoes from Banff were assigned to escort Dallachy Beaufighters to Flekke Fiord, a short way up the Norwegian coast from Lister. Flight Lieutenant Bob McColl of 455 Squadron led again. He had with him Flying Officer Jack MacDonald, from Melbourne. Before the War Jack had worked as a commercial artist. He joined the militia with the Army Service Corps. In May 1940, aged 18 years, he enlisted in the Second AIF and saw service in Palestine, Egypt, Libya,

Greece, Crete, Syria and Lebanon as a driver with the 2/106 Australian General Transport Company. Jack returned to Australia in 1942 and joined the Air Force for pilot training in Tasmania, Canada (including a General Reconnaissance course) and Britain, with the Empire Air Training Scheme.

This day, 11 January, landfall was near the southern tip of Norway just east of Lister Light. McColl approached Lister at 1500 feet. Just below them, at 1000 feet, the German fighters were waiting. There seemed to be seven Messerschmitt Bf109s and several Focke-Wulf Fw190s, and crews saw at least one other German aircraft take off from the local air base. The twenty-one Beaufighters kept close together and flew north-east past Lister. A section of the Mosquitoes took on the Messerschmitts and Focke-Wulfs.

McColl turned towards the sea and set course for base. He sent this radio message, 'Attacked by fighters - target not attacked'. At this stage crews saw two Mosquitoes chasing a German fighter that was pouring smoke, and which soon dived into the sea. In the end, three Messerschmitts and one Focke-Wulf were shot down for the loss of one Mosquito, one Beaufighter, and an Air Sea Rescue Warwick which was last seen investigating the area where the Mosquito had crashed.

This operation prompted the Commander-in-Chief of Coastal Command to ask for a second Mustang fighter squadron to protect his strike wings in Scotland. Shortly after, 19 Squadron RAF was ordered to join 65 Squadron RAF at Peterhead.[7]

Weather for the remainder of January was bad, on occasions atrocious. The Daily Mail newspaper for Wednesday 24 January 1945 reported,

> Snow, which held up trains and buses for hours...isolated Scottish villages; WAAFs at an RAF Coastal Command station in the North-East of Scotland saved the aerodrome from becoming isolated in a five day battle during the 50 mph gale.[8]

455 Squadron navigator Jack Tucker noted that, on 26 January, it was snowing and twelve degrees below freezing. It snowed for days and all Squadron members were called to shovel snow from the runways.[9] The snow finally began to thaw on 30 January. Colin Milson wrote to his father from Dallachy at the end of January 1945,

> We have just had a week or so of heavy snow falls and now it is all thawing so conditions are rather unpleasant. Unfortunately, we all have to dig and keep digging and shovelling the snow off the runways and it is not very pleasant around about midnight either.
>
> The news from Europe is extra good - looks as though old Joe [Stalin] is really determined to get to Berlin in a very short while. .. .I don't fly very much these days - lots of office work to do and anyhow I've almost finished my tour - actually I could have finished several months ago but 'Ops' and the squadron life appeals to me so I have been taking things pretty easily over the last four months.... Got a couple of jackaroos in the squadron and we have quite a few yarns about old times...Cheers and love to all at home. Affectionately, Woodge.[10]

Despite the terrible weather the strike wings at Dallachy and Banff made seven attacks along the Norwegian coast during January 1945. 455 Squadron were away on the 6th, 8th, 9th, 10th, 11th and 12th. After that, the weather was far too bad and apart from a few unsuccessful patrols, they were locked in at Dallachy for almost a month, until Friday 9 February 1945.

At last, early on that Friday, two New Zealand Beaufighter crews on reconnaissance from Dallachy found a German naval force deep inside Forde Fiord, inland from Ytteroerne Light. Several merchant ships had also been sighted in a neighbouring fiord, but the Air Officer Commanding 18 Group, Air Vice-Marshal Ellwood, decided to attack the naval target. A strike from the Dallachy Wing was ordered and that afternoon thirty-two Beaufighters from 144, 404, 455 and 489 Squadrons began to gather in formation over Dallachy. Fifteen minutes later they set course for Peterhead to join twelve Mustangs of 65 Squadron RAF, their fighter escort, and the force turned for the Norwegian coast.

This operation was the first for Flying Officer Harold Spink and his navigator, Flying Officer Lloyd Clifford. Spink had joined the Empire Air Training Scheme early during the War. To give himself a better chance of

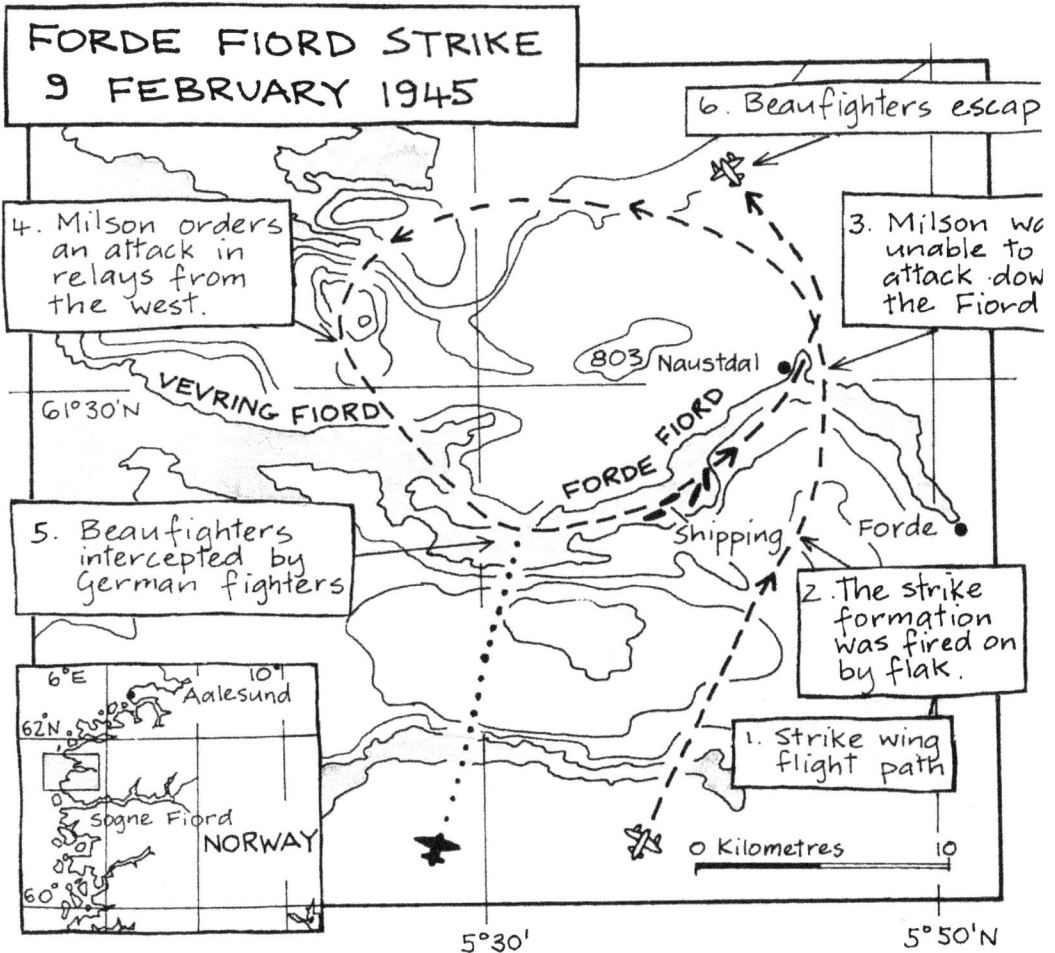

FORDE FIORD STRIKE
9 FEBRUARY 1945

6. Beaufighters escap

3. Milson wa
unable to
attack down
the Fiord

4. Milson orders
an attack in
relays from
the west.

61°30'N

VEVRING FIORD

803 Naustdal

FORDE FIORD

5. Beaufighters
intercepted by
German fighters

Shipping

Forde

2. The strike
formation
was fired on
by flak.

6°E   10°
Aalesund

62°N

Sogne Fiord
NORWAY

60°

1. Strike wing
flight path

0 Kilometres   10

5°30'   5°50'N

being enlisted into the Air Force he had started private flying lessons. His flying skills led to his employment as a flying instructor when he arrived in Britain, and it wasn't until 16 January 1945 that he and navigator Clifford joined their first operational unit – 455 Squadron.

Wing Commander Colin Milson led the Dallachy force and they made landfall near Utvaer Light off Sogne Fiord at 3.40 pm, an hour and a half after leaving base. The formation swept north and flew up the coast at eight hundred feet. Visibility was from thirteen to sixteen kilometres with a cloud base at two thousand feet. The Norwegian mountains were covered in snow and the fiords were nearly white with floating ice. Jack Tucker recalled, 'Our Outriders reported "no target" but when we flew [north] over Forde Fiord we were fired on by very accurate flak.' It indicated a target.

Milson saw many ships grouped together halfway along the Fiord amongst floating ice, some close to shore and others in the middle of the Fiord. The

'Brilliant leadership, gallantry and careful planning': The tough and charismatic Commanding Officer of 455 Squadron at Dallachy, Squadron Leader (later Wing Commander) Colin Milson DSO and Bar, DFC and Bar, who took over command from Wing Commander Jack Davenport in October 1944, as 455 Squadron moved to Dallachy in Scotland.

main vessel was a heavily armed *Narvik* class destroyer protected by several escorts.[11] The *Narvik* class destroyer was en route from northern Norway where it had been part of a destroyer force. It was attended by two tugs. Milson ordered an attack down the Fiord, from the east and towards the sea, but the destroyer and its escorts were anchored close to the land and were in a position difficult to attack from the east. Milson called off this attack, then turned to fly west towards the mouth of the Fiord. He briefed the Beaufighters for an attack in relays from the west, this time up the Fiord. His port sections were to attack vessels on the port side of the Fiord and starboard sections to attack their side. Milson led 455 Squadron in a climb, then dived in to attack the larger escort and a minesweeper with rockets and cannon.

The ships were well defended. The flak from positions on the cliffs and from the ships formed a crossfire – Flying Officer Jack MacDonald has strong recollections of the flak defences here. He remembers,

> a virtual wall of 20 and 37 millimetre tracers into which 455 had to attack. It would be the nearest thing to World War 1 'Going over the Top', then required of all Beaufighter and Mosquito strike squadrons in Coastal Command.[12]

Harold Spink was close to Milson, in the first wave, and on his leader's port side. Spink says,

> As we flew down the Fiord I started taking hits from the flak. I could see it hitting my wings, coming closer to the cabin. Then as the flak hit the cabin I fired my rockets. My arm was hit and paralysed. I pulled out of the Fiord and my navigator, Lloyd Clifford, came forward to help me. There was a smell of hydraulic fluid and I was having trouble controlling the starboard engine.[13]

Flying Officer Bill Herbert was following. His navigator, Jack Tucker, recalls,

> My pilot, Bill, selected an armed escort vessel for our target...The flak did not abate. In those split seconds we were waiting to be hit - there seemed no alternative. When it came we were not surprised. There was a dull thump and Bill warned, 'prepare to bail out'. He had good reason because his control column became loose in his hands, at about 100 feet above the ice on the Fiord water.[14]

The Australians were followed up this Fiord by Beaufighters from 404 Squadron RCAF and 144 Squadron RAF which attacked the destroyer, scoring numerous strikes. Crews saw a violent explosion on the bridge and saw debris fly into the air. The large escort was by then smoking heavily with flames visible amidships.

Two squadrons of Focke-Wulf fighters from *Jagdgeschwader* 5 had been scrambled from their base near Bergen just after the Beaufighters crossed the Norwegian coast. A dozen of these fighters came in from the west, up Forde Fiord, and they intercepted the Beaufighters during the strike.

The flak and German fighters now took a heavy toll. Two Beaufighters were seen to crash into the hills with engines on fire. A Beaufighter was hit by fire from a shore battery that it was attacking and crashed near its target, and another Beaufighter was seen to crash on a hill after a mid-air explosion.

'Tearing towards the end of the Fiord at dot feet with faulty elevators': Navigator-Wireless Operator Flight Sergeant Jack Tucker (left) with his pilot Flying Officer Bill Herbert, both from Melbourne, who operated Beaufighters with 455 Squadron from October 1944 until the end of the War in May 1945. Their aircraft was hit by flak during a strike on shipping in Forde Fiord on 9 February 1945. The flak severed an elevator control wire just behind the pilot's seat, and Herbert flew with great skill to manoeuvre out of the steep sided Fiord before piloting his aircraft back to Dallachy.

Flight Lieutenant Bob McColl of 455 Squadron attacked the destroyer. He was himself attacked by a Focke-Wulf which made four separate passes at him; then he was hit by a shore-based flak battery. Two other Australians, Warrant Officer Donald Mutimer of East Brunswick, Victoria, and his temporary navigator Flight Sergeant John Blackshaw of Penshurst, New South Wales, died when their plane was shot down.[15]

Where they could, local Norwegians went to the aid of the aircrew. A Beaufighter of 144 Squadron RAF landed in the water near a farm. The farmer's wife saw it tearing through the trees towards the Fiord. She called to her husband and son who ran to their jetty and rowed out to the sinking aircraft. The two crew members were pulled from the water and carried up to the farmhouse. The navigator was injured and needed medical attention and to get a doctor the Germans had to be notified. The two men, Holly and his pilot, Smith, became prisoners of war.

The Beaufighters now turned for home. Harold Spink's Beaufighter had lost a starboard window and he was wounded in the chest, ribs and right arm. He was attempting to get his damaged aircraft home, with a broken compass, an unserviceable intercom and the pitch control on one engine stuck at 2700 revs. Harold says,

I was unable to talk [to my navigator]. There was a small notice on the dashboard which said "Warning, Modified Tail Plane". I pointed to the [individual] letters on that sign to spell out messages to Lloyd...Lloyd put on a tourniquet and gave me a shot of morphine. Our radio was gone but Lloyd gave me a course for base.[16]

Bill Herbert also struggled with a damaged plane. His navigator, Jack Tucker, says,

Very soon Bill had the plane under control. He found that he could raise the nose of the aircraft but couldn't lower it. The plane was not on fire but we were tearing towards the end of the fiord at dot feet with faulty elevators and with Fw190s buzzing around.

With great skill and airmanship Bill manoeuvred UB-F out of the fiords and over the sea. He found that by winding on nose heavy trim he could keep the aircraft in level flight by pulling back on the control column...When we were a reasonable distance from the coast the next job was to trace the extent of the damage. I decided to check this from the elevators forward. Behind my cockpit was a hinged armour plate door. I stripped off my harness, Mae West and flak helmet and squeezed through the small opening. I crawled to the very end of the fuselage and discovered that two steel wires controlled the elevator. I gently pulled on one and it came out freely. The other was firmer. I then retraced my crawling until I discovered the damage a couple of inches behind Bill's seat. There was a neat 20mm diameter hole in the port wall of the cockpit. It had severed the loose wire but just grazed the other. Close![17]

It was more than two and a quarter hours later before crews made it back to Dallachy. Bill Herbert was the first of 455 Squadron to touch down, with Flying Officer Chick Smith. Next, ten minutes later, came Warrant Officer Noel Turner, Flying Officer Jack Cox and Flying Officer Harold Spink, who says,

As we got near Dallachy we saw they were firing Very pistols because of some problems on the field. I had my undercarriage down but it wouldn't lock. I came in to land on the grass verge next to the runway and the wheels folded up as we hit the ground. Next, the propellors began to hit the ground. I pressed the foam button to make sure the engines didn't catch fire. As we came to a stop, the Commanding Officer of the New Zealand Squadron, Wing Commander Dinsdale, was there to pull me out of the cockpit.[18]

Finally, for 455, came MacDonald, Furlong, Thompson, and the leader, Milson.

Harold Spink was taken to nearby Elgin hospital and he was attended by a Russian orthopaedic surgeon. He was a walking patient when the War ended and he rejoined 455 Squadron at the RAAF Holding Unit at Beccles. Spink and Clifford were both awarded the Distinguished Flying Cross. Milson's recommendation for Spink reads, 'This officer displayed great courage in pressing home the attack after being hit and very commendable fortitude whilst on the way to base. Having a complete starboard side window of his cockpit blown out, rain was pouring in and the bitter cold constituted an added trial.'[19]

'The Beaufighter often proved its ruggedness': Beaufighter LZ 407 was taken on charge by 455 Squadron on 24 March 1944. It was hit by flak and bullets on four separate occasions, the last during a strike at Aardals Fiord on 4 April 1945, when it was badly damaged by flak but courageously flown back to Sumburgh in the Shetland Islands by Flying Officer Steve Sykes. It is shown here undergoing repairs at Dallachy after a strike at Forde Fiord on 9 February 1945. (Photograph courtesy J. Tucker)

404 Squadron RCAF lost six of their eleven Beaufighters on the strike. Of the twelve crew members of 404 Squadron shot down only one (Flying Officer Roger Savard) survived. That night at Dallachy, 455 Squadron's Medical Officer, Flight Lieutenant Bob Macbeth wrote,

> Tonight the wind is whistling around the Nissen huts and among the pine trees of our wood. The Spey is rushing down, filled with the freezing water of melting snow. A few blocks of ice are still tumbling in the current. This afternoon, Bob Mc [McColl] went missing. 9 [aircraft] lost from station altogether - my hut is empty now - only Frank across the wall is stoking his fire - Johnnie Pilcher is on leave...all so different from the flame and flak and destruction in the fiord this afternoon when Bob was killed. I can hardly believe it. So clearly can I see his smile and feel the power of his grip, as I congratulated him on his engagement to Katie three days ago. It was his last trip. Seems funny to think that he too, will never grow old.[20]

The day after this attack, Colin Milson wrote to Bob McColl's mother, advising her that her son was missing and that, 'Owing to the rapid sequence of events, it was impossible to say what happened to his aircraft.' Mrs McColl wrote back to Milson, 'We have not given up thinking we will see him again, though things may not point that way now.'[21]

It was then several weeks after this attack that Bob Macbeth wrote to McColl's friend, Flight Lieutenant David Whishaw.

> It was a bit grim here the day Bob [McColl] went and I felt pretty poorly about it. We lost two that day. Mutimer and Blackshaw were the other crew...They

went in on a destroyer and a lot of flak support craft. Spink got some flak in his Rt arm and came home with it stuck up on the gun sight. He did a belly landing with the red lights of empty tanks showing. I think the Canadians lost 6 because they were the last in. Most went to the [German] fighters...It seems that Bob [McColl] pressed in on the destroyer and some say he went round twice - anyway I don't think there is really any hope and we have not yet heard of any picked up. There should be some of course, but they think Bob went straight in...Katie is remaining at work and does not want to go on leave.[22]

Katie Ivatt was Bob McColl's fiancee; they had become engaged shortly before this strike and Bob had taken her for a flight in a Beaufighter, in the name of LAC Ivatt, just six days before this operation. Convinced of his loss, McColl's companions sold his car, a 1936 Ford, in nearby Elgin, though it was reported that 'from all accounts it was in a dilapidated state of repair.'

Miraculously, Bob McColl and navigator MacDonald were not killed. MacDonald had been able to give instructions to his pilot as they were being attacked by the German fighters, and they avoided the fire from the Focke-Wulf. But then they were hit by flak from a shore battery. McColl took his damaged Beaufighter west, towards the mouth of the Fiord and the sea. After some twenty-five to thirty kilometres, close to the coast, he was forced to ditch. They escaped from their Beaufighter unhurt and spent thirty minutes in their aircraft dinghy before being picked up by a Norwegian fishing boat, escorted by a German naval vessel. They spent six days in an Oslo jail and were then taken to Stalag Luft 1 in Germany on 8 April 1945.[23] At the camp there were several other Beaufighter aircrew shot down on that strike at Forde Fiord, including Pilot Officer Smith, 144 Squadron RAF, shot down by a Focke-Wulf. Smith had lost an engine and was unable to clear the Fiord on half power. He ditched and was picked up by the Norwegians after three hours. Another was Flying Officer Roger Savard, of 404 Squadron RCAF. He had crash-landed on the solid ice at the top end of the Fiord after being attacked by the German fighters. Savard was slightly injured but his navigator was killed.

After being searched at Stalag Luft 1, McColl was handed over to the British personnel at the camp. He described the camp routine in his diary,

We have two roll calls per day, 8.25 and 4.10, and actually these are welcome because they split up the day and are something to do. We have to go into the barracks immediately the air raid siren goes and cannot go outside or even open the windows. Two chaps were shot some weeks ago for going outside...The days are mostly the same but we visit other huts for a cup of something between meals. There is plenty of tea and coffee in the Red +[Cross] parcels, the only difficulty being fuel which is scarce...This routine continued on 'til the 30th April when the 'Ruskies' were only a few miles away and all the Jerry guards bolted west to hand themselves over to our armies rather than be caught by the Reds. The first Russians came in at midnight on 1st May and next day pandemonium broke out. Some drunken officer arrived and compelled our CO to release the boys, who once started went wild...

On the 2nd May parties were dispatched to the local airfield to get it into condition for our evacuation. We were all very optimistic and expected to get out in a few days and spirits were very high.[24]

By 10th May, there were no signs of transport aircraft. They were now in Russian occupied territory and the men were frustrated by the rumours of an early trip back home, and by missing the VE Day celebrations. Bob McColl finishes his diary, 'It certainly is disheartening but by disbelieving all rumours and thinking of the joys waiting for me when I do get out I manage to keep a more or less happy interior. Another week or so though and I think I will be feeling somewhat sorry for myself.' During his time in captivity, McColl wrote these words,

> For all my comrades on the line tonight
> Out there among the crags and tempestry
> I breathe a soldier's prayer, begrudging me
> This billet where there is little light
> My empty useless shoulder is contrite
> With warmth and life, although my door be barred
> My straw is paradise but sleep is hard
>
> With knowledge that I share not in the plight
> We who are prisoners do not complain
> For we are sheltered from the stormy places.
> But there are some tonight who face the rain
> And some who do not feel it on their faces
> Oh, what's the use of shelter overhead
> If on my heart I feel the rain instead.

Finally, on 13 May 1945, Bob McColl was flown in an American Fortress aircraft back to England. Katie Ivatt recalled that time when Bob finally reappeared,

> V.E. Day came and went and on May 13th I was summoned to the M.T. [motor transport] Office. Wandering vaguely what I'd done wrong, I went in and a very excited Corporal told me Bob was on the phone. It was the happiest and most wonderful day of my life. I remember my mother telling me how Bob rang my home first when he got back [from] Germany. My father answered and she heard him say, 'Oh my dear boy' and she knew who it was.
>
> I met Bob at my parents' home in Surrey when he returned to England. I got compassionate leave the morning he rang me and went home by train...I walked from the station on the road to home and could see him standing at the gate waiting and we just raced to each other. We were married 3 weeks later and Bob was recalled from leave on our honeymoon.[25]

The attack at Forde Fiord had resulted in the loss of nine Beaufighters. This was a setback for 18 Group and added weight to the Group Commander's plea for another squadron of Mustang fighters to escort his strike wings. 19 Squadron RAF joined 65 Squadron RAF at Peterhead only weeks later.

In the weeks following the battle at Forde Fiord, the bodies of eleven airmen were recovered and buried in unmarked graves in the churchyard at Forde. On 17 May 1945, Norway's national day, the local people gathered at the churchyard for an official ceremony. A priest performed the funeral rites and the graves were decorated with flowers and twigs of spruce. Beside each grave stood a girl in national dress holding a wreath of flowers. Next year, the eleven graves were moved to a war cemetery in Haugesund: three men remained missing.

Jack Tucker and his wife returned to Forde Fiord in 1976. They took a taxi to the Fiord and found that their driver had witnessed the strike as a sixteen year old boy. Jack says that the Norwegians have an enormous respect for the airmen who lost their lives in Forde Fiord and he was very moved by their feelings. In the 1980s interest began to revive in Norway's involvement in the War. The editor of a local newspaper appealed for a memorial to the memory of the fourteen men who died at Forde Fiord. The idea was taken up by the local branch of the Reserve Officers Association and the Forde community contributed generously to a fund. On the 8th of May 1985, the fortieth anniversary of VE Day, a stone monument was unveiled bearing the names of the Canadians, Britons and Australians who lost their lives at Forde Fiord. Jack Davenport also visited the site in 1987, and he laid a wreath at the memorial.

Three 455 Squadron crews joined crews from the Canadian 404 Squadron and New Zealand 489 Squadron crews for a roving sweep along the Norwegian coast on 26 February. It was early evening when they arrived off the southern tip of Norway, just south of the Lindesnes Light, and they set course easterly coastwise. They flew on past Ryvingen Light off Mandal, and then received a radio message advising that there was a target off Lindesnes. They continued east a little further, then circled and headed back along the coast. Off Mandal they found and attacked a merchantman, an escort vessel and a tanker, then broke off. The merchant ship was loaded with mines and a moment after being torpedoed it exploded spectacularly in all directions, fragments hitting some of the escaping aircraft.

On the way back to Scotland, Flying Officer Joe Brock's Beaufighter developed engine trouble. Brock ordered his navigator, Flight Sergeant Bill Hirst to bale out. Hirst landed safely near the railway station at Inverurie, north-west of Aberdeen. Brock's Beaufighter was seen to roll over and he bailed out near the coast north of Peterhead, in Aberdeenshire. Brock's parachute opened and he was carried towards the sea by an easterly wind. Three witnesses followed on their bicycles but they were unable to find the pilot.[26] An extensive land and sea search by members of the Squadron was undertaken at Peterhead for Flying Officer Brock, but he could not be found and he was presumed to have drowned.

This was another tragic loss for the Squadron; Joe Brock's younger brother Jim had been killed with 455 Squadron on an operation near Heligoland just six months before, on 10 August 1944. Joe Brock's body was later recovered, and he was buried in a churchyard at Hune in Denmark, twenty-eight kilometres north-west of Aalborg.

During February 1945, anti–shipping tactics continued to develop. The Admiralty established a running plot of all German shipping and convoy movements along the Norwegian coast which was compiled from photographic reconnaissance, reports from agents, and an analysis of radio traffic from the ships themselves. This plot was a great success, not only for the strike wings but also for planning attacks by Royal Navy torpedo boats, submarines and the Fleet Air Arm. Coastal Command also proposed attacks on lighthouses down the Norwegian coast, to make it more difficult for the Inner Leads to be used for night–running of convoys. After lengthy negotiations with the Admiralty and the Norwegian Government in exile, these attacks were authorized from the start of March. By the end of March, some fifteen lighthouses would be out of action, keeping vessels at anchorage in fiords for long periods while they waited for weather and sea conditions to be suitable for sailing without the usual land beacons.

Early in March 1945, Warrant Officer Peter Ilbery and Warrant Officer Bill Mitchell were told to report to Coastal Command's 18 Group Headquarters at Pitreavie Castle. They were to be interviewed by the Air Officer Commanding Coastal Command, Air Marshal Sir Sholto Douglas. Both were to be considered for an officer's commission. Peter says,

> Bill and I should have gone from Dallachy together, but at the last minute another crew was needed for an operation and he was selected. I farewelled Bill down at 'flights' and arranged to meet him later at Group as I had ten days leave. I drove alone down Loch Ness then to Edinburgh and finally to Pitreavie Castle.[27]

A Wing strike was ordered that day, 8 March. All four Dallachy Squadrons had Beaufighters on the patrol and the force of twenty-nine aircraft was led by Squadron Leader John Pilcher of 455 Squadron. Pilcher's pilots included Warrant Officer Bill Mitchell, with eleven operational sorties behind him including the successful strike against the *Sirius* in Fuglsaet Fiord on 9 January.

Two Outriders separated from the main force two hours after take–off and they turned inland south of Bremanger, making for Midgulen Fiord. It was a return visit to Midgulen for Dallachy - they had attacked several merchant vessels there in early November 1944, just after the move from Langham. This time they found six ships tucked in tightly against the cliffs at the top of the Fiord, three on each side. The geography here was no ally for the attackers - the Fiord sides were sharp and steep, and as Pilcher led in he had to put his Beaufighter into a steep dive. At the top of this Fiord, and first in line for the attackers, was the 2960 ton merchant ship *Moshill*. This ship had been built in 1940 as an oil burning fruit carrier and passenger ship. Its movements suggested that it was being used to carry troops and military stores. It had been seen in the Baltic Sea, in Copenhagen, and had left Kristiansand near the southern tip of Norway just three days earlier. A little further down this fiord was the 3655 ton merchant ship *Ex-Larenburg*, also known as the *Alsterdam*. It had been laid down for the Hillersberg Steamship Company in Holland before the War and was first seen in air photographs when being fitted out in Rotterdam in June 1940. The *Ex-Larenburg* had been last seen at Kristiansand on 17 January 1945.

'He had to dive hard down the steep cliffs': Flight Lieutenant Neil Smith fires a rocket salvo at a merchant ship in Midtgulen Fiord on 8 March 1945. Twenty-nine Beaufighters attacked six ships that were found tucked in against the steep sides of this Fiord.

Pilcher attacked with cannon and rockets. The Beaufighters were up against flak from the ships, from camouflaged land positions on both sides of the Fiord, and from the town of Midtgulen. Over the ships, white puffs of exploding flak filled the sky. Flight Lieutenant Neil Smith followed Pilcher. Smith recalls that he was flying low along the hills surrounding the Fiord and he had to dive hard down the steep cliffs. Immediately below him, very close to the cliffs, he saw the *Moshill* and the *Ex-Larenburg*. When his cannon shells were striking around the *Moshill* Smith fired his rockets and then pulled away sharply to fly away down the Fiord. Behind Neil Smith came Flying Officer Chick Smith DFC, and he saw Neil's rockets slam into the *Moshill*. Then came Warrant Officer George Longland who also put a well directed rocket salvo into the ship which was by this time enveloped in smoke. Beaufighters from 404 Squadron RCAF attacked a car ferry on the opposite side of the Fiord. The Beaufighters flew out past Bremanger and set course for base. The *Moshill*, the *Ex-Larenburg* and the car ferry were all badly damaged.

Some twenty-five kilometres from the coast of Norway, the leader of the escorting fighters from 65 Squadron RAF saw red tracer arcing down to the sea and he spotted two aircraft travelling fast towards the Norwegian coast, two partly inflated dinghies in the sea and some wreckage, but no sign of survivors. One crew from 455 Squadron, Warrant Officer Bill Mitchell and his navigator, Flight Sergeant Ivor Jury failed to return to Dallachy and it was presumed that they had been shot down and killed by enemy fighters. Later, Flying Officer Jack Cox reported that he had been following Mitchell into the attack and that Mitchell had not been shot down during the strike.[28]

'He always pressed home his attack': Flying Officer (later Squadron Leader) Neil Smith from Melbourne who flew Beaufighters with 455 Squadron from March 1944 until May 1945. He was awarded the Distinguished Flying Cross in May 1945 for skilful leadership, efficiency and courage.

Several days later, Peter Ilbery eventually arrived at Pitreavie Castle to see Sholto Douglas, and was shown the underground operations room. He recalls,

> I saw details of Bill's operation that occurred several days earlier. He was posted as missing. On the way back to Dallachy I told my navigator that Bill Mitchell was missing. My navigator told me that he had known but hadn't known how to tell me. The formation had been jumped by Fw190s and Bill's Beaufighter had been shot down. Bill had always said that he wouldn't be killed.[29]

There was now a period of two weeks before the Australians had another strike. For much of that time the weather was impossible along the coast of Norway, so the Dallachy squadrons trained in formation flying, in rocket practice and in exercises in joining up with their fighter escorts. Several times they attempted to reach Norway but were hampered by bad weather and poor visibility. There was an aborted strike on 23 March, when they were unable to confirm their location and lost formation in the confusion. They tried again the next day.

Dallachy crews gathered at their aircraft in mid–afternoon on Sunday 24 March. At 3.45 pm Flight Sergeant Eric Nayda and his pilot, Warrant Officer George Longland, checked their equipment. Nayda opened his navigator's log, 'Preflight Equipment OK'.

Squadron Leader John Pilcher led Longland and six other Australian crews away at 4 pm. The force leader was Squadron Leader Christison of 404 Squadron RCAF. Christison circled the Dallachy field while a force of thirty-eight Beaufighters, fourteen Mustang fighters and two air-sea rescue Warwicks made formation. Nayda recorded in his log that he set course for Norway at 4.22 pm. They were at one hundred feet above the sea and 460 kilometres from their planned landfall at the Egero lighthouse, on the south Norwegian coast. Estimated time of arrival was 5.58 pm. Nayda set the cannons and his rear gun to Fire. Exactly thirty minutes later, he calculated his position and recalculated the distance and time to landfall.

Only minutes before 6 pm they finally arrived at the Norwegian coast. The

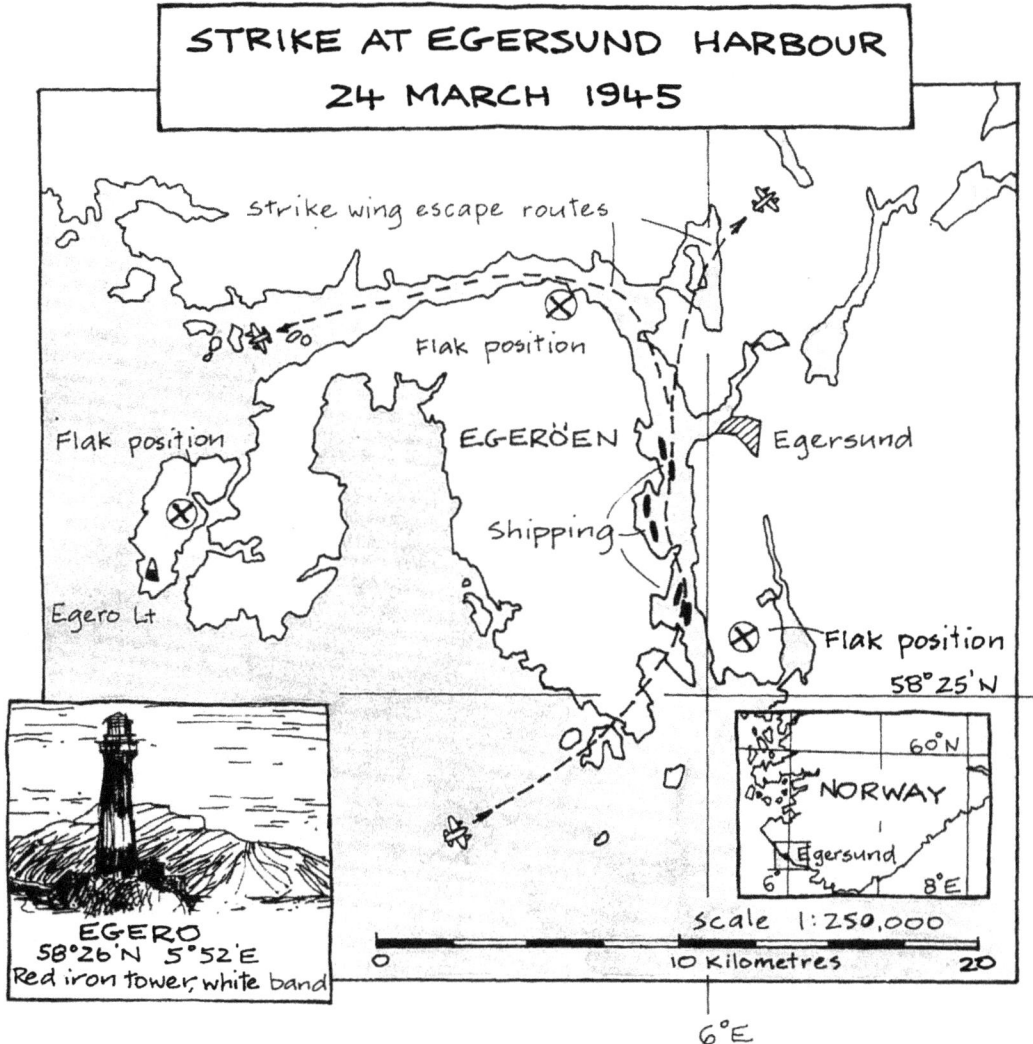

STRIKE AT EGERSUND HARBOUR
24 MARCH 1945

Strike wing escape routes

Flak position

Flak position

EGERÖEN

Egersund

Shipping

Egero Lt

Flak position
58° 25' N

EGERO
58° 26' N 5° 52' E
Red iron tower, white band

60° N

NORWAY

Egersund

6°   8° E

Scale 1 : 250,000

0   10 Kilometres   20

6° E

Outriders had been there and reported to Pilcher the position of several merchant ships in the outer harbour at Egersund. There was no cloud but the weather was very hazy; visibility was down to three kilometres at the coast. The Beaufighters came in from the south-west, climbing to an attacking height of some 2000 feet, then over the southern tip of Egeroen Island and into the passage that forms the outer harbour at Egersund.

Christison led down onto the merchant ships. He strafed with cannon then let his rockets go. The other Beaufighters followed close behind. The Harbour was a dangerous place; flak defences were intense and accurate but the force pressed their attack on the merchant ships at top speed. There were puffs of flak above the harbour sky and Beaufighters were swarming over the ships. During the attack Longland's Beaufighter was seen to crash into the surrounding hills and explode.

Flying Officer Bill Herbert and navigator Flight Sergeant Jack Tucker were on the port side of the Beaufighter formation. They attacked, then continued north up the outer harbour, and soon after Herbert banked west following the channel towards the sea. Tucker was watching out through the side of his navigator's cupola as they wheeled to port around the tip of Egeröen Island at the top of the outer harbour. They swept around low and close to Egeröen

'Beaufighters were swarming over the ships': Flying Officer Lloyd Clifford took this photograph facing the rear in his Beaufighter navigator's cupola during a strike on enemy shipping at Egersund Harbour on the Norwegian coast on 24 March 1945. One 455 Squadron Beaufighter was shot down during this attack.

Island and Tucker saw German anti-aircraft gun crews running towards their weapon on his port side, just a short distance away. He reached for his rear gun but the gun crew were quickly out of his range. Flight Sergeant Peter Ilbery was also on the strike. He says,

> Apart from the incredible flak my memory of that attack is how forcefully it was pressed up against the vessels and then the jinking and yawing almost on the water and ground on the way out to be under the anti-aircraft fire. Scraping up a slight hill overlooking the harbour there came the sight of a small group of Norwegians gesticulating deliriously.[30]

The Beaufighters escaped in several directions and heavy flak bursts followed them out to sea. The attackers had been bold enough to venture into the defended harbour and Coastal Command described it as a 'daring strike'.[31] But losses had mounted - two Beaufighters exploded in the strike area, one dived into the sea and another (Longland) crashed on land, killing the crew. An aircraft from the Canadian squadron ditched twenty kilometres off Egeröen, and the force leader, Christison, ditched some twenty-five kilometres south-east of Egeröen.[32] Norwegian authorities reached the wreck of the only Beaufighter that crashed on land – UB-X flown by Warrant Officer George Longland. The Norwegians recovered the log sheet that Errol Nayda kept during his flight across the North Sea to Egersund. That log sheet is now held in the archives at the Australian War Memorial.[33]

In the last week of March 1945 Eisenhower was compiling his plans for the Allies final campaign in the West. Although the German Army had been decisively defeated in the field, he believed that there would be no surrender so long as Hitler lived. He felt that they would have to destroy the remnants of the German Army piece by piece, with the final possibility of a prolonged campaign in the rugged alpine area of western Austria and southern Bavaria known as the National Redoubt. It was also widely believed that Hitler would continue the fight from Norway.

In March 1945 the Mosquitoes of the Banff Wing gained increased range, by using drop tanks, and were able to penetrate into the Skagerrak and Kattegat to attack at sea and on the eastern coast of Norway as far as Oslo. On March 30, forty-four Mosquitoes found shipping alongside the quay at Porsgrunn and attacked four merchant vessels and a barge. One aircraft, flown by Squadron Leader Philip Davenport, the brother of Wing Commander Jack Davenport who was the former Commanding Officer of 455 Squadron, was shot down. It was Jack Davenport who had planned this strike, in his capacity as operations controller for 18 Group's strike wings. Philip landed his Mosquito on a frozen lake, and was captured by the Germans.

On 31 March came an official announcement of the end of the Empire Air Training Scheme. Because very few RAAF aircrew had arrived in Britain in the preceding nine months there was little impact on RAAF units overseas. In Australia it produced a large surplus of semi-trained aircrew and instructors. In Britain, the carefully trained aircrew in 455 Squadron still had important work ahead.

9

Strikes in Procession

Small wonder that the Commander of the German Fighter Force remarked ruefully that the fighting had degenerated into a 'Wild West show put on by the Allies.'

Air Chief Marshal Tedder, Marshal of Royal Air Force

I put a tourniquet on his left arm, plugged the worst wound and gave him a shot of morphine. He made no complaint, although it was bitterly cold, and he was determined to get back to base.

Flying Officer Vic Pearson describing the injury to his pilot after a strike at Songe Fiord on 4 April 1945.

At the briefing we were told, 'Gentlemen, we expect heavy losses'. I had my foot on the lower rung of the aircraft when transport drove up and said the War was over and the strike was off.

Pilot Officer Jack Tucker, 455 Squadron Navigator

APRIL was a month of anticipation; it seemed that Germany was certain to be defeated. The Allies were pressing on Germany from both sides and the possibility that Germany's new air and naval weapons might delay the verdict had been destroyed by Allied bombing. But the decision by the Three Powers to persist in the demand for unconditional surrender also made it seem certain that the War would not end until the Western and Eastern armies met in Germany.

While part of the RAF was concentrating on helping Montgomery's armies in their advance into Germany, Coastal Command's strike wings continued to pursue German shipping sheltering in the Norwegian fiords. The strike wings began to benefit from the improving weather, from a continued build-up of equipment and experienced crews, and from the efforts of Germany to move shipping and stores back home. Lord Tedder, Marshal of the Royal Air Force wrote, 'Small wonder that the Commander of the German Fighter Force remarked ruefully that the fighting had degenerated into a "Wild West show put on by the Allies"'.[1]

The Dallachy Wing planned a return in force to Egersund Harbour on 3 April 45. They rendezvoused off Peterhead with the Mustang fighter escort of 19 Squadron RAF before setting course for Norway. Near Egero Light Milson wheeled them to starboard. Here the weather was very poor, a complete cloud cover and rain showers kept visibility down to several kilometres. Then heavy flak opened from near Egero. Pilots struggled to keep station in this large formation of thirty-five Beaufighters; they had little height, were under fire, and manoeuvring in rain became hazardous. Pilots knew they had to keep their formation tight but the risk of collision was high. Milson wheeled again to starboard and away from the coast and ordered the force to return to base. Soon after a VHF radio transmission was heard, 'aircraft going to ditch'. The Beaufighter flown by Warrant Officer Tom

455 Squadron aircrew at RAF Dallachy. From left to right: F/Sgt Jack Tucker (navigator), W/O Tommy Furlong (pilot), W/O Ivor Gordon (navigator) and F/Lt Stuart Addison (pilot). W/O Furlong was killed when he was forced to ditch his Beaufighter while returning from a patrol to the Norwegian coast on 3 April 1945.

Furlong was seen flying low over the water. Then it hit and bounced, with the port engine trailing white smoke and starboard undercarriage down. It hit the water again and disintegrated. The aircraft dinghy inflated when the aircraft settled in the water and aircrew saw a swimmer in the water making for the dinghy. Another body was seen in the water.

The main formation circled the area of the ditching until Milson ordered them to set course for Dallachy. Flying Officer Tom Higgins remained on station in the area in his Beaufighter until the Air Sea Rescue Warwick arrived. Furlong's navigator, Flight Sergeant Williams was rescued but Tom Furlong was never found.

455's Jack Cox was flying Outrider again on 4 April; he and navigator Ibbotson were very skilful at this now. They took off ahead of the force from Dallachy and were looking for a ship reported to be unloading at Aardal, a small port at the extreme eastern end of Sogne Fiord some 160 kilometres inland. Allan Ibbotson said their job was to,

> accelerate ahead [of the main attacking force] as we neared the target to confirm it was 'as reported'- and if not to search nearby anchorages to locate it. Upon sighting, the pilot of that aircraft would head toward the strike force and his navigator would fire a Very pistol using the colours of the day for identification and then proceed to lead the formation into the attack from the most appropriate direction. Being unarmed, the Outrider would break off at about 1200 yards from the target...to observe and take photographs.[2]

Some crews recall that at their briefing it had been suggested that a ship unloading at Aardal had been carrying 'heavy water' for use in the manufacture of atomic weapons. This day, Cox found the target, reported it and guided the strike force in. There were eight rocket-equipped aircraft flown by 455 Squadron and six cannon-firing aircraft from 489 Squadron as anti-flak escort. The 455 Squadron section was led by Flying Officer Steve Sykes.

As Sykes flew through the entrance to Sogne Fiord his plane was hit by shore based flak batteries. Despite the danger he stayed in formation, led on up the long Fiord then wheeled north-east up Lyster Fiord. Finally, they bore around to the south-west to approach Aardal along a lake. The target was a 3000 ton steamer riding high in the water and lying very close to a concrete wharf. Ship and shore batteries put up a heavy barrage as the strike aircraft attacked. John Ayliffe was also an Outrider that day, but he managed to join the attack. He says, 'The Fiord at Aardal is very narrow and is surrounded with mountains 1400 feet high. The attack had to be carried out with aircraft diving down in line astern'.[3] Although wounded, Sykes stayed to lead the attack and he was hit again, suffering a compound fracture to his left arm and flak wounds to one leg. His navigator, Vic Pearson said, 'Flak burst at the port side close to the pilot's window and not only was he sprayed by shrapnel but also by small pieces of perspex from the window.'[4]

With difficulty, Sykes pulled his plane out of the dive, away from the steep sides of the Fiord and turned for home. Pearson crawled forward from his navigator's compartment to help. He says,

I put a tourniquet on his left arm and plugged the worst wound and gave him a shot of morphia. He made no complaint, although it was bitterly cold, and he was determined to get back to base. Standing behind him, I did my best to steer the Beaufighter, but he had to work the rudder himself, although his left leg was well peppered.[5]

As Outrider, Jack Cox and Allan Ibbotson watched the attack and saw several hits on the target above and below the waterline. Then they turned for home. Ibbotson saw a Beaufighter to starboard, with the cockpit light on.

I received an Aldis signal flashed from this aircraft [from Pearson] that his pilot was injured. I was instructed by my pilot to let them know that we would escort them; and was also instructed to radio a coded message to base stating that we were escorting an aircraft with an injured pilot. Soon after I received a coded message to alter to 'low' frequency and broadcast frequently to enable the airborne Warwick aircraft to home-in on us so that they could release their airborne dinghy, should a ditching become necessary. We were also instructed to alter course and head for Sumburgh, which was a much closer landfall for the troubled crew.[6]

All the way back to Sumburgh, Pearson stayed in the cockpit with his pilot and tried to keep him warm by holding his own flying jacket over Sykes' shoulders. As they prepared to land, Pearson worked the hydraulic controls and engine throttles for his pilot. Sykes made an excellent one-handed belly-landing. Their escort, Jack Cox, had to circle Sumburgh for about thirty minutes while Sykes was taken from his plane and the aircraft cleared from the runway. Sykes had dozens of shrapnel fragments to his arms, legs and body and several fractures above the elbow of his left arm. For his action, Sykes was recommended for the Immediate Award of the Distinguished Flying Cross. Pearson was also recommended for the Distinguished Flying Cross. His recommendation reads,

He held himself responsible for guiding and keeping his pilot in formation with the escort aircraft although it was a dark night...This officer, since losing his original pilot, F/O Collaery, has flown on operations with 8 different pilots of this unit. His keenness, capability and willingness to fly with any pilot has been an outstanding feature of his operational work on this squadron.[7]

That was Steve Sykes' last operation for the War. He stayed at Sumburgh until 5 May when a 455 Squadron aircraft flew him back to Dallachy. By then he had recovered well from his wounds and he was sent to a RAF rehabilitation centre. Some final assessments are from fellow aircrew; Noel Turner says, 'Sogne Fiord was one of the longest, nearly cutting the country in half and it must have been a tough trip back for Steve and Vic Pearson.'[8] David Whishaw says that Sykes was 'a quite exceptional air fighter, pure and simple'.[9]

Next day, on 5 April 1945, eight 455 Squadron Beaufighters accompanied seventeen other aircraft from Dallachy on a shipping strike off Hareid Island. From landfall at Sandoy Light, near Sogne Fiord, they flew on up the

Norwegian coast. They looked into Vevring Fiord at the western end of Forde Fiord, and into the eastern end of Midtgulen Fiord. They crossed Nord Fiord, flew north-east to Syvde Fiord, over Rovde Fiord and then finally on to Hareid Island just south of Aalesund. The leader heard over the VHF radio, 'Bandits at 6 o'clock'. Six Messerschmitts, with greeny-grey fuselage and a drop tanks amidships, were flying on a parallel course and at the same height as the Dallachy Wing.

The leader, Squadron Leader John Pilcher, ordered his formation to throttle back to 190 knots and to close up. Two Messerschmitts attacked the Beaufighter flown by Flight Lieutenant Keith 'Bluey' Moore. Warrant Officer John Ayliffe was also there in the formation, flying low over the water. His navigator warned him of the enemy fighters. Ayliffe says, 'I took evasive action but witnessed fighters attacking two Beaufighters on my port side. These Beaufighters were flying in formation. One was literally shot to pieces and before crashing, collided with the other plane resulting in both crashing into the sea.'[10] Moore had collided with the Beaufighter flown by Flying Officer Geoff Hassell from Hyde Park, South Australia. In the fight that followed, the escorting Mustangs shot down four enemy aircraft. Both Beaufighter crews were lost; Flight Lieutenant Keith Moore, from Coogee, New South Wales, his navigator, Flight Sergeant Ron Bull, Flying Officer Geoffrey Hassell and Flying Officer Ted Loonam from Maroubra, New South Wales.

During time off, airmen from Dallachy would go to the local town of Fochabers to visit hotels, or travel further to Elgin, Inverness, or Edinburgh. They also invited local girls to dances at Dallachy. A young lady who lived in the nearby town of Buckie in those days was Isabel McKay Stewart. She remembers that 'My mother was always very, very sorry for the lads of Dallachy and made any we brought home very welcome'.[11] Isabel later published a book of recollections, called 'En Kin Ee Mine Es?' [and can you remember this]. She writes about these war times,

> The lads that flew the Beaufighters that ye mine or noo,
> Jist young loons wie great courage, steadfastly time,
> Money a nicht ye'd go awa tae meet yer date,
> Bit fin he'd been straffin Norway, he'd hae met his fate.

On 7 April, twenty-one Beaufighters from Dallachy attacked two merchant vessels, an escort vessel and a tug in Vadheim Fiord, off the northern side of Sogne Fiord. Flying Officer Jack Cox was Outrider again. He was hampered by low cloud along the Norwegian coast but found the target and circled for ten minutes waiting for the main body, sending reports to the leader.

The main body included six crews from 455 Squadron, including Warrant Officer Ian Murray and his navigator Flight Sergeant Don Mitchell. Murray was from Tambo in Central Queensland and before the War he had been working as a jackeroo. In 1941 he left the farm to go to Brisbane to join the Navy. With an old school mate he stopped off in Toowoomba and was having a few beers in a hotel when they were engaged by an RAAF recruiting officer. Murray decided to join the RAAF and he learned Morse code and navigation by correspondence (the local post master tested him for Morse). He was finally called up early in 1942. After training and some time spent in 66 Squadron RAAF, flying anti-submarine and convoy controls along the Queensland coast, he arrived in England just before D-Day in 1944.

The strike force made the Norwegian coast at Utvaer Light and turned inland. They tracked to Vadheim, a small town at the northern end of Vadheim Fiord, off Sogne Fiord. The approach was from the north-east; and there was the four thousand ton merchant vessel *Oldenburg* at Vadheim Quay. She had been built in 1914 at Wesermunde. During the War *Oldenburg* had been seen constantly on the Norwegian coast and in German ports, and it was believed that her refrigerated cargo space had been carrying fish from Norway to Germany. She was fitted with defensive armaments in Bremerhaven in the summer of 1944 and had been last seen in Hamburg in February 1945. One hundred metres south of *Oldenburg* and close to the cliffs was another merchant vessel of some three thousand tons.

It was bright here, with the sun reflecting off the water in the harbour as the Beaufighters sped in low over Vadheim. Crews could see clearly the two storey buildings around a small town square. The Beaufighters first attacked the *Oldenburg* at the quay and followed on to the other ships. A well directed salvo of rockets from Flying Officer Tom Higgins hit the water just short of the *Oldenburg* near her starboard quarter - a wet hit. This ship was seen smoking amidships, and soon the other merchant ship was almost totally enveloped in smoke. Another ship further away from the quay, an auxiliary, was almost obscured by smoke. The Beaufighters broke away out of Vadheim Fiord and then back into Sogne Fiord.

On the way down Sogne Fiord, Warrant Officer Ian Murray's aircraft was intercepted by a flight of six Focke-Wulf Fw190s. The German fighters chased Murray around the Fiord for ten minutes, attacking him one at a time. Murray's navigator, Mitchell, fought back with his rear gun, damaging at least one attacker. Mitchell recalled,

> My pilot hugged the sides of the mountains and fiords, dodging in and out, with the 190s keeping close at us. We were so close in that the fighters had to keep one eye of the cliffs and the other on their sights to fire at us - which did not improve their aim. As each one came in and fired he would break away and then join the line coming after. We must have doubled back on our tracks several times.[12]

Murray finally eluded the fighters and made for Dallachy. He says,

> Fortunately it was the one area of Norway I knew fairly well so we didn't go up any dead-end fiords. When the navigator told me they [the Focke-Wulfs] were all heading around the opposite side of a mountain, I pulled out into a small cloud out to sea. It was only then that I found I had no instrument panel.[13]

Murray's radio transmitter was working, but not his receiver.

When I looked back at my navigator his place was filled up with a tangle of trailing aerials. It seems the Air-Sea Rescue Warwicks were able to hear that we were heading for the Shetland Islands and followed us across. Unfortunately the drome there was under about 50 feet of fog.

Murray decided to follow the Scottish coast down to Dallachy. Flying Officer Howard of 455 Squadron flew his Beaufighter out to meet him, but when Howard pulled alongside and Murray tried to throttle back to a normal landing circuit speed, his aircraft began to shudder so he abandoned the escort.

I had no hydraulics, hence no wheels or flaps. A 20mm shell had blown out the hydraulic tank about 18 inches from my head! Our elevator trim wires had been cut so it was quite an effort to hold it up at slow speed. We made one normal approach but started to overshoot to hell so I went around again, this time getting right on the deck out to sea and coming in over the land at about 10 feet, throttling right off and holding off the deck at about three feet. We stalled and slid in without a shudder. I fired off both engine fire extinguishers. Next thing my nav had me by the throat trying to pull me out with my harness still secure! I was still collecting my maps. Col Milson was about the first to arrive as he jumped off the control tower top floor and ran.[14]

The Allies were now overrunning Germany. By 11 April US forces approached the Elbe River and RAF fighters and fighter bombers swept ahead of the advancing ground columns to attack enemy movement against negligible opposition from the Luftwaffe.

'Fortunately we didn't go up any dead-end fiords': Warrant Officer Ian Murray (left) and his navigator Flight Sergeant Don Mitchell were part of a formation which attacked shipping in Vadheim Fiord on 7 April 1945. After the strike their Beaufighter was intercepted by six German Focke-Wulf fighters. Murray was chased around the fiords by the fighter aircraft for ten minutes, taking several hits, before he escaped into a small cloud and then out to sea.

That day, Coastal Command Intelligence reported armed merchantmen, their escorts and at least one minesweeper in Fede Fiord, just north of Lister. Standby crews at Dallachy were sent for an early lunch before the pre-strike briefing at the base operations complex. The strike force was twenty Beaufighters from the Dallachy Wing, with an escort of Mustangs and air-sea rescue Warwicks. After briefings, trucks and 'twitchers', carried the aircrew to their squadron dispersal areas around the airfield. 'Twitchers' were small Hillman utilities with a canopy over the back and were given that name by crews to describe the state of anxiety that invariably built up during preparations for a major strike over Norway. Flying Officer Peter Ilbery and his navigator, Flying Officer Bill Bawden, prepared for take-off in their aircraft, T-Tommie,

> From synchronised watches each squadron had a start up time. The noise of the engines being run up was stark contrast to the wireless silence we'd maintained. A green Very signal showed that the operation was on and the first Squadron rolled to the end of the runway.

At the end of the runway, crews did their final checks; navigators recited the list to pilots. Hatches closed, Hydraulics OK, Trim for take-off, Mixture rich, Air Ok, Pitch fully fine, Fuel OK, Flaps set one-third for take-off, Gills open, Gyro set, Navigator OK. Pilots lined up in their Vics of three. Peter Ilbery continues,

> I checked our momentum briefly, turning on to the right of the runway to hold the aircraft on its brakes. The release. The twin Hercules engines advanced quickly and smoothly and Tommie burst down the right side of the runway, buffeted by Johnnie's W, one hundred yards ahead on the left. Doubtless Bob's N, staggered to the left on the runway was partly in my slipstream. Throttle forward to the gate then through it to give the maximum boost for lift-off. Then, with the roughness in the controls subsiding as the undercart came up followed by fifteen degrees of flaps, airborne and smooth riding to join the other aircraft at eight hundred feet.

Once in formation over Dallachy, the three Beaufighter squadrons followed the coastline of Banffshire to the east and were joined by a dozen Mustangs and two Warwicks. The formations dropped down to the water, flying at fifty feet to avoid German coastal radar. Peter Ilbery recalls,

> My station on the extreme right of the formation meant there was little for pilot or navigator to do as the essential flying height above wave top was set by the lead pilot. Landfall computation involving wind, airspeed and G-box tuning was done by his navigator ...Too much time to anticipate the approaching action and wishing it were over...My wing tucked in behind 'Bakers' wing put the navigator, Tuck, facing backwards in his armed cupola, level with me. Occasional acknowledgement made across thirty feet of airspace in breaks from routine tasks like keeping an eye on gauges measuring oil temperatures and pressures, cylinder head temperatures and fuel...The nav was asked yet again to estimate how long to go to landfall.

In these trips from Scotland to Norway, Beaufighter crews were occupied with repetitive tasks. Pilots flew to maintain station, checked instruments, radio altimeter, airspeed and compass. The navigator listened out for calls from base and plotted the course. As Norway came nearer, both scanned their sectors of the sky for enemy fighters and the coastline for an indication of the exact point of arrival at the coast.

The coast appeared and the wing climbed. They made landfall just west of Lindesnes and climbed to just a few hundred feet above the bare Norwegian mountains. They flew on north for less than ten minutes and wheeled south. Almost immediately the ravine of Fede Fiord opened out below them, a long stretch of calm water, hazy in the distance, with many steep headlands jutting into the Fiord. The formation tightened as it continued its turn to face down the Fiord. There was a quay at the town of Lervigen and a 2500 ton escort ship was moored there, with a long building behind the quay and several merchant ships nearby. Peter Ilbery says,

> Being on the extreme inside of the turn had necessitated throttling right back and my speed became dangerously low and it felt as if the Beaufighter in nose up attitude was struggling not to drop out of the sky. Nothing for it but to throttle on and I commenced to ease out in front before the command 'attack-attack-attack' came from the leader. Black puffs of exploding heavy flak started to appear. The minesweeper was clearly edged against the side drop of the fiord. Over on the left in a group of vessels an armed merchantman was in my line of attack. The anti-aircraft fire came nearer. With the attack command now given it was a relief to be able to push the throttles through the gate and regain full control over the aircraft. I was now well out in front to be the first in to run the gauntlet.

Peter Ilbery's aircraft was now directed down at the merchantman which was broad-side on against a rocky promontory. A mile of water was between him and his target.

> Cine camera on, safety catches off and then through the din of the four cannon pounding beneath my feet actually heard that puff explode. It was a very near miss. My cannon hits raked the cliff above the vessel, so pushed the nose down a little more and then a single staccato crack beneath the aircraft like the noise accompanying lightning. Smell of fireworks inside the aircraft, port motor smoking but the controls still responding.

Ilbery was still too far from his target to let the rockets go normally. With a damaged engine he could only ease the nose up slightly and fire his rockets, hoping to loop them in. Then he took evasive action.

> The movements to avoid remaining an easy target were then instinctively performed - jink sideways, down, sideways, up and repeat, repeat. Tommie responding to the rough treatment of rudder bar and stick but there remained the question of the engine.

He was faced with the decision of whether to shut down his port engine or, with smoke abating, wait. Although the engine was running roughly, the thought of returning all the way home over the North Sea on one engine led

'Grinding and pounding across the grass': Flying Officer Peter Ilbery (right) and his navigator Flying Officer Bill Bawden, at the 455 Squadron base at Dallachy. They survived a belly-landing in their damaged Beaufighter after a strike against shipping in Fede Fiord on 11 April 1945.

him to wait and see. Ilbery cut his revs right back. After a few parting shots from the coastline the engines were eased back further to conserve fuel.

> Informed nav what was happening and he tried to talk to any other aircraft. We had straggled and although having communications no other aircraft in sight. Keep broadcasting on the Warwick's wavelength. With the sun setting low on the horizon in the direction of Scotland, the eyes smarted, the back ached and the mind whirled.

'Tommie' was the last aircraft back to base and Ilbery asked permission to land.

> Undercart selected...no familiar drag of wheels going down and no locked wheel indicator lights. Talked to base and flew over the control tower...wheels confirmed not down. I was told to fire off any remaining ammunition and use up any remaining fuel. Flew out over the darkening sea...We made a long flat approach across the countryside. I selected flaps and realized it would be a fast contact with the ground. We just cleared the perimeter fence and floated down to feet, then inches above the ground. Still did not settle. Made a touch of stick forward; impact! Grinding and pounding across the grass surface with dust spewing into the cockpit. Came to rest with the fire truck and station commander's car coming alongside.[15]

Peter Ilbery opened his canopy and dropped onto the wing as navigator Bill Bawden escaped from under his cupola. It was the last flight for this T-Tommie.

That evening the United States Ninth Army reached the Elbe and next day secured a bridgehead across the River. Now, the Americans were only eighty-five kilometres from Berlin and the Russians were advancing at a similar distance to the east.

Jossing Fiord, some twenty-five kilometres south-west of Egersund Harbour and near the Norwegian town of Sogndal, was well known for an incident in February 1940. Then, the German supply ship *Altmark* had been making its way south along the Norwegian coast with nearly three hundred British merchant seamen captured by the German battleship *Graf Spee*. Norway was neutral and Britain argued that the passage of prisoners of war through Norway's territorial waters was a breach of neutrality. Germany assured Norway that the *Altmark* carried no prisoners. Not satisfied, the British ordered HMS *Cossack* to intervene. *Altmark* made for the safety of Jossing Fiord. *Cossack* followed, came alongside *Altmark*, boarded, and after a struggle collected the prisoners and departed the Fiord. Although *Altmark* had run aground during the boarding, it was refloated, returned to Kiel, renamed *Uckermark* and re-employed as a supply ship.

Wing Commander Colin Milson led twenty Beaufighters into Jossing Fiord on 14 April. They found five ships near the top end of the Fiord; there was an M-Class minesweeper tied up at the zinc factory wharf, and outside it a U-boat of some 500 tons: a little further up another M-Class minesweeper. Near the very top of the Fiord was a tanker of over 5000 tons, and the depot ship *Adolf Ludritz* tied up at Titania Quay. The *Adolf Ludritz* had been built at the beginning of the war as an E-boat depot ship and had been based at Egersund before moving to Jossing Fiord. The tanker had been followed on air photographs since early 1942, and it was suspected that it was a Norwegian tanker that had been captured after the German invasion. It had been sighted often in Kiel, Hamburg and Bremerhaven. The sides of the Fiord were very steep and rocky, with a narrow road along the northern side which continued on past the Fiord and zig-zagged up the hill at the end of the harbour. All the ships were moored close to the steep cliffs of the Fiord.

Milson led in and the aircraft seemed to take the defences by surprise. Milson was on the U-boat quickly, and with Warrant Officer Ian Murray close behind they both took it on. It was a steep dive onto the ships, and at 260 knots they had only a moment after release of their rockets to watch them in flight. Milson reported that,'Crews following us in said we got hits and they could only fire at the centre of the cloud and smoke covering the U-boat and the naval auxiliary.'[16] Flight Lieutenant Clive Thompson took the next target after Milson's, the second M-Class minesweeper. In his path the defenders had fired two rockets that trailed heavy wires to ensnare the Beaufighters, the white rocket trails clear against the rock of the surrounding cliffs. With his cannons belting the water around his target Thompson let his

'They had to abandon formation and attack in procession': A Beaufighter flown by Flight Lieutenant Clive 'Tommy' Thompson DFC, fires rockets at a merchant ship in Jossing Fiord on 14 April 1945. This photograph also shows the wire-trailing rockets, fired by the defenders. Flight Lieutenant Thompson and his navigator Warrant Officer Ivor Gordon remained near the Fiord for an hour after this attack to keep watch over a 489 Squadron RNZAF crew who ditched their aircraft.

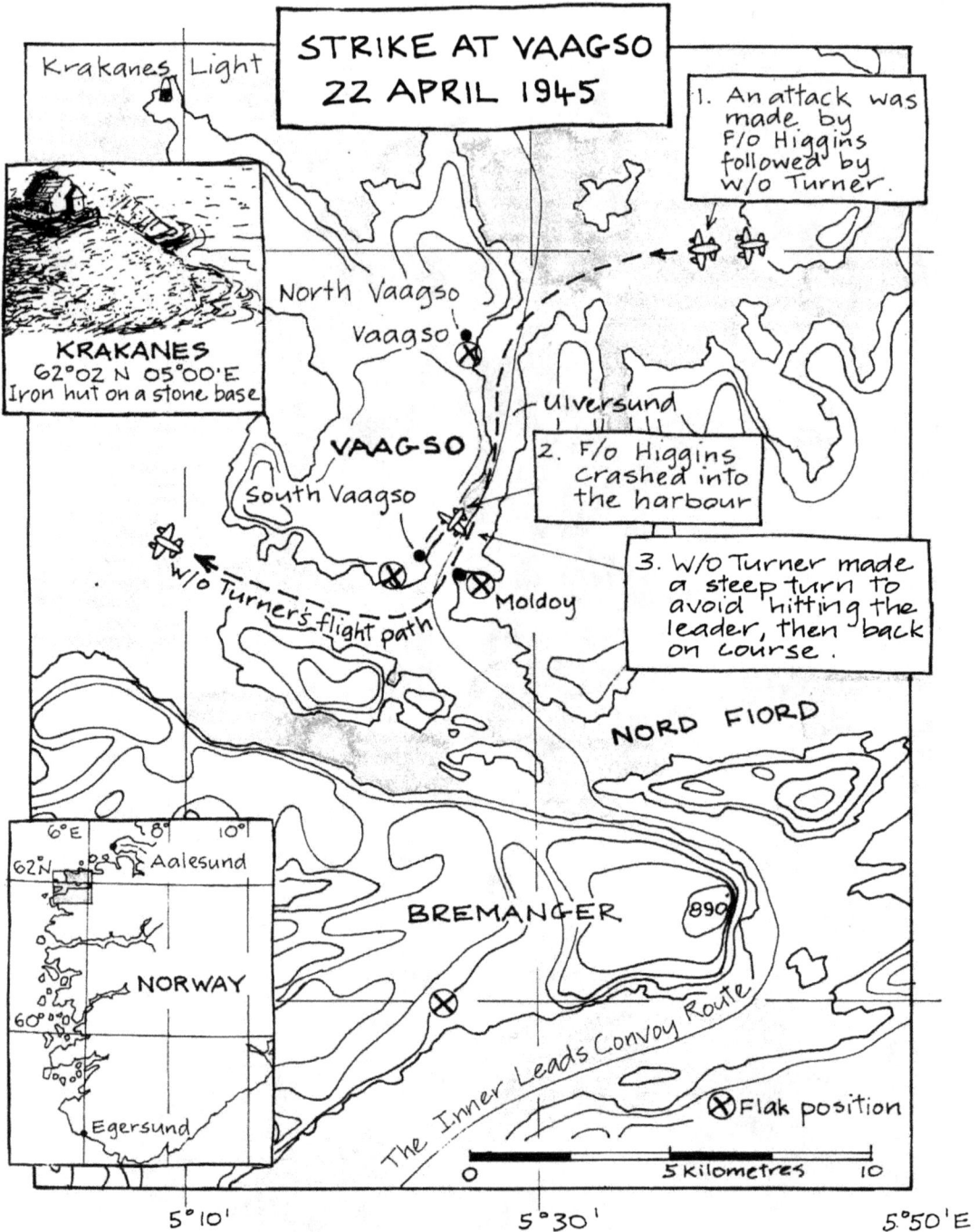

STRIKE AT VAAGSO 22 APRIL 1945

Krakanes Light

KRAKANES
62°02 N 05°00'E
Iron hut on a stone base

1. An attack was made by F/o Higgins followed by W/o Turner.

North Vaagso
Vaagso

Ulversund

VAAGSO

South Vaagso

2. F/o Higgins crashed into the harbour

W/o Turner's flight path

Moldoy

3. W/o Turner made a steep turn to avoid hitting the leader, then back on course.

NORD FIORD

6°E   8°   10°
62°N   Aalesund

NORWAY

60°

Egersund

BREMANGER   890

The Inner Leads Convoy Route

⊗ Flak position

0   5 kilometres   10

5°10'   5°30'   5°50'E

salvo of eight rockets go, then pulled up and over the cliffs near the end of the harbour. The others followed up the Fiord which was so narrow that they had to abandon their formation and attack in procession. Beaufighters were converging on the targets, flying dangerously close to each other, and there was the added danger from the flak being fired from both sides of the Fiord and from the escort ships.

Two 489 Squadron aircraft collided on their way out of the Fiord; one crew ditched and made it into their dinghy about fifteen kilometres from the coast. Milson and his deputy leader Thompson stayed with the stranded New Zealanders. Although Thompson's port engine was damaged he flew with Milson over the dinghy for an hour, homed in the ASR Warwicks and waited until an airborne lifeboat was dropped before he set course for base.

On 22 April it was nearly all over for Hitler. Russian artillery shells began to fall in the garden of the German Chancellery, a German counter attack against the Russians south of Berlin had failed and Russian troops were in the northern suburbs of the capital. Hitler was in despair. He decided that, rather than go to Berchtesgaden as planned, he would stay in Berlin and direct its defence in person. If this failed, Hitler said he would shoot himself. Hitler seemed to have accepted that he could no longer continue.

On this day, Flying Officer Jack Cox was on a reconnaissance sortie in the vicinity of the island of Vaagso, near Krakanes Light. He recalls,

> I discovered four ships under way in an open fiord just to the north of Vaagso Harbour. We radioed their position and returned to base. On arrival at Dallachy I found a strike force under W/Cdr Gadd had been briefed and was almost ready to take off.[17]

Jack Cox was immediately refuelled and sent out again as Outrider for the force. He says that, 'This meant that by the time we reached the target area two hours would have elapsed since the sighting, and as the ships had been under way at the time they would have moved into a safe anchorage.' At Vaagso in 1941, Lord Louis Mountbatten had launched the first of three operations that secured the reputations of the Commandos (the others at St Nazaire and Dieppe). On Christmas Eve 1941, a naval cruiser, four destroyers and two infantry assault ships escorting soldiers of 3 Commando had attacked the towns of South Vaagso and nearby Moldoy.

Now, 18 Group formed another large strike formation from Dallachy, twenty-five Beaufighters, escorted by two Mustang squadrons (19 and 65 from Peterhead – twenty-four fighters). There were six ships in Vaagso Fiord, including four merchant vessels. Jack Cox says,

> I led the strike force inland and saw the ships anchored against the docks at Vaagso. I gave their position. Vaagso Fiord had high precipitous sides, is very narrow and the docks are on a narrow strip of land at the base of the western cliffs. The attack from east to west was suicidal, particularly without preliminary anti-flak attack.[18]

The Beaufighters flew to an area north-east of Vaagso, and considered the approach. The Squadron Operations Book records that 'it was too late to deploy force so leader told such aircraft as could attack to do so'. Flying Officer Tom Higgins of 455 Squadron wheeled in, followed by Warrant Officer Noel Turner. It was a long approach down the length of the Fiord, known as Ulversund, with a flat sea, high clouds and almost unlimited visibility. There were flak defences near North Vaagso, Moldoy and South Vaagso, and bright tracer was thick around the harbour. Noel Turner says,

'I began violent evasive action and couldn't seem to stop': Warrant Officer Noel Turner fires a rocket salvo at shipping near the town of South Vaagso on 22 April 1945. The lead Beaufighter in this attack, flown by Flying Officer Tom Higgins with his navigator, Warrant Officer Alan Mirow, was shot down and both were killed.

'At the entrance to the Fiord weather was clear, visibility good, I could see a ship at the far end. I could see no reason for not attacking.' They had flown into intense flak defences. Higgins attacked the merchantman alongside the quay and followed up with an attack on a warehouse. His aircraft was seen to crash in the sea near an escort ship. Turner carried on the attack; he was also down close to the water with spray from his cannon shells partly obscuring the target when he finally released his rockets. Turner recalls,

> F/O Higgins had disappeared half way through the attack; he reappeared, heading slowly toward my gun sight. Apparently he had dived steeply and made a low level flat approach...I stopped firing, swerved steeply away, then started again. As we pulled up over the target I began violent evasive action and couldn't seem to stop. Eventually I settled down much to the relief of Bert Iggulden who must have wondered if he had to navigate while jinking and jagging all the way back to base.[19]

Following this attack, Noel Turner was recommended for the Immediate Award of the Distinguished Flying Cross. His Commanding Officer wrote that 'This pilot has always displayed an outstanding fighting spirit and may be depended upon to press home each attack to the bitter end'.

Flying Officer Tom Higgins and his navigator, Warrant Officer Alan

Mirow, were killed. They had completed twenty-nine operational sorties together in the eight months they had flown with 455 Squadron.[20] This operation was the third time that Jack Cox had been Outrider that month. On a recommendation for the Distinguished Flying Cross, Wing Commander Milson wrote of Cox, 'He is a most successful recce pilot and has very intimate knowledge of the Norwegian coast. His Outriding has been outstanding and on two attacks he was absolutely invaluable to the leader of the strike force.'

Late next morning, on 23 April 1945, Flight Lieutenant Clive 'Tommy' Thompson led a wing strike against four ships lying along the west side of Risnes Fiord, running south just inside the entrance to Sogne Fiord. It was the first time Thompson had led a wing strike and according to Thompson's navigator, Warrant Officer Ivor Gordon,

> I assumed that Tommy was selected to lead the strike to spread the responsibility around and Tommy was a pretty experienced pilot. Colin Milson the Commanding Officer flew number two and I remember looking back into the barrels of his cannons the whole way across to Norway.[21]

There were twenty Beaufighters from Dallachy on the strike, six from 455 Squadron. As usual they were escorted by Mustangs of 19 and 65 Squadrons RAF and accompanied by air-sea rescue Warwicks. They crossed the Norwegian coast up by Holmengra Light and Thompson tracked north-east towards the familiar Sogne Fiord. At Risnes Fiord, they found shipping tucked in very close to the steep sides of the Fiord. First was a whaler type auxiliary, and next the 4969 ton merchant vessel *Ingerseks*. The *Ingerseks* had been built in 1914 and acquired by her Norwegian owners in 1943. After being taken under German control she had been working cargoes of coal, mainly from North Sea ports to Norway. The next, a merchant ship of some 1000 tons, had been used for coastal trade, and the last was a small auxiliary.

Thompson brought them in over steep cliffs from the south, heading towards Sogne Fiord. The huge, steep cliffs of Risnes Fiord dwarfed the ships and the aircraft that attacked them. The Beaufighters dived down in succession and some pilots found themselves cramped for room when they attempted to line up the ships. Thompson attacked, followed by Milson who made a successful cannon run at the whaler auxiliary, setting it on fire. Then in came Cristofani who strafed flak positions on the shore. In the lead aircraft, Warrant Officer Ivor Gordon recalled,

> During the attack Colin Milson was so close to us that he didn't seem to fire a single shot for fear of hitting us. I thought that was very good of him. The cannons of the Beaufighter had a habit of setting fire to the insulation surrounding the cannon. After the attack Tommy asked if I was smoking a cigarette. I told him I wasn't and I found a fire taking hold near the cannon boxes. I emptied the fire extinguisher onto it which quietened the fire down but I had to finish the job with my hands.[22]

The Squadron Operations Record Book records Gordon's trial, 'The fire had reached the wing roots and the pilot prepared to ditch, but by this time navigator had beaten out the fire'.[23] As a result of the attack the German

'The huge cliffs dwarfed the ships and aircraft': Warrant Officer Albert Vigor of 455 Squadron took this photograph through the rear of his navigator's cupola as his pilot, Warrant Officer John Ayliffe, flew down Risnes Fiord after attacking shipping on 23 April 1945.

merchant vessel *Ingeseks* of nearly five thousand tons was left listing and on fire. For this action and for other outstanding work, Ivor Gordon was recommended for the Immediate Award of the Distinguished Flying Cross. Clive Thompson was also recommended for the Immediate Award of the Bar to his DFC.

Flying Officer Bill Edwards from Sydney was tasked to lead a 455 Squadron section on a patrol to Fede Fiord three days later, on 26 April. They left Dallachy in the early afternoon, crossed the Norwegian coast south of Egersund, and swept east across Lunde Lake. Beaufighters wheeled around to starboard to approach Fede Fiord from the north-east. Outriders from 489 Squadron had seen shipping in Brevigen Bay in Fede Fiord. They made the report, 'three Merchant Vessels end of Fiord'. The ships were lying very close to the high cliffs. The *Palmyra* was a 3600 ton coal burning merchant vessel which had been launched in 1943 and had been seen in the Bergen area towards the end of 1944. After visiting northern Norway, she had been seen again around Bergen.

Wing Commander Tony Gadd from 144 Squadron RAF led the Beaufighters into the Fiord. Just fifteen days earlier the Dallachy Wing had attacked shipping in the same Fiord. One 455 Squadron crew, Pilot Officer Bob Penhall from Mascot, New South Wales and his navigator Flying Officer Eric Blakers from South Perth were now on their second trip to this place. Brevigen Bay sheltered two merchant vessels and two escorts. From this direction only the merchant vessel *Palmyra* and one of the the escorts were exposed for attack; the other two were sheltered by cliffs. As the Beaufighters came in to attack along the cliffs on the north side of the Bay a wire–trailing rocket was launched into the air by one of the escorts, right across the path of the aircraft closing in on the *Palmyra*. Flak came from the escorts and from at least four gun batteries on the land. 455's Flying Officer Wallis attacked the *Palmyra*, scoring some very accurate cannon hits. A Beaufighter from 144 Squadron RAF was hit and crashed into the water at the entrance of the Bay and near the *Palmyra*.

Emerging from the attack they were intercepted by a force of up to ten Focke-Wulfs and Messerschmitts to the north of Flekke Fiord. In 455 Squadron's lead aircraft, Edwards' navigator Warrant Officer Keith Hamilton of North Mackay, Queensland, radioed that his pilot was wounded but thought he could make base and he asked for an escort. Almost immediately Hamilton called that he was being attacked by fighters. There was no further word from him; his aircraft was seen to crash and explode on a hill between Fede Fiord and Flekke Fiord.

Warrant Officer Ian Murray was also attacked by two Messerschmitts; he had survived a long attack by enemy fighters several weeks earlier and this time he was rescued when the escorting Mustangs from 19 Squadron arrived. This was the last attack in Norwegian waters by 455 Squadron. One ship was sunk and another damaged. Keith Hamilton and Bill Edwards were the last operational casualties suffered by 455 Squadron in the War. In the period between 5 and 26 April, 455 Squadron had taken part in seven anti-shipping strikes and fourteen reconnaissance patrols.

At the beginning of April 1945 the Allies suspected that the Germans might withdraw into the mountains of Bohemia and lower Bavaria to make a final stand. Montgomery's and Bradley's Army Groups had both crossed the Rhine in the third week of March and they continued to seek out and destroy the German Army. The Russians launched their attack on Berlin on 16 April with two and a half million men and by 22 April the Russians were fighting in the streets of the German capital.

In the last week in April, no shipping of any type was found in the lower half of the North Sea, and there was virtually no movement seen on the Norwegian coast. Intelligence revealed that Germany was suffering severe fuel shortages and the merchant marine had its oil allocation reduced by up to thirty percent.

In the early hours of 1 May, Hitler's Army Chief of Staff, General Krebs, told the Russians that Hitler had been dead for two days and asked the Russians for an armistice. The Russians refused to negotiate and demanded

'A wire-trailing rocket was launched into the air': The Dallachy Strike Wing attacked shipping in *Fede Fiord on 26 April 1945. The Beaufighters faced flak from escorts and shore batteries, and rock-* *ets that trailed steel wires that were launched across their path. One aircraft from 455 Squadron,* *flown by Flying Officer Bill Edwards, was hit by flak and then shot down by German fighters. This* *was the last strike in Norwegian waters by 455 Squadron.*

that the Germans unconditionally surrender on all fronts. Krebs went back to the centre of Berlin and shot himself. The next day, 2 May 45, the German commander of the Berlin garrison capitulated. Admiral Doenitz, President and Commander-in-Chief of the Armed Forces after Hitler's death sent a representative to Montgomery to negotiate a surrender in the west. Montgomery also refused to negotiate.

On 3 May 45 there was a rush move of aircraft, crews and groundstaff from Dallachy south to RAF Thornaby, better to attack shipping attempting to escape from Germany. Nine 455 Beaufighters were away from the Dallachy base before midday and another two left later to join the Squadron at Thornaby. Just before 5 pm, Wing Commander Colin Milson led a force of Beaufighters and Mustangs from Thornaby for Kiel Bay. Twenty-six Beaufighters made formation over nearby Huntcliffe Point at 5.15 pm and set course to a point on the western coast of the Jutland Peninsula. Cannons were set to 'Fire', fuel switches were set to the outer wing tanks, and at five hundred feet and 180 knots the crews settled in to the routine for the flight to their target area.

Flight Sergeant Bill Waldock was navigating for Flying Officer Bob Cristofani. After the operations briefings Waldock had marked his map, showing the planned flight path taking them across central Jutland, then down the eastern coast of Odense and finally west-south-west back to Langham. Two hours from Thornaby they arrived at the Jutland Peninsula and took a slight turn to starboard. North-west of the island of Odense the formation wheeled south, heading for the waters around the approaches to Kiel. Now it was just after 7 pm and they began their patrol, with a heavy cloud cover making conditions grey and dull. Just off the west coast of the island of Langeland, at the southern end of Langeland Channel, they found seven ships in line astern heading north, including three merchant ships. There were also several Tank Landing Craft of the type built in considerable numbers for the German invasion, but never used for that. These were, according to Jack Cox, 'crammed with German soldiers.'[24] The formation swept across the convoy from the port side, 455 concentrating on two minesweepers. Cannon raked the hulls and crews reported multiple strikes from their rockets. At least four of the attacking aircraft strafed the landing craft and the soldiers fired back with their rifles. Jack says, 'I hate to think of the plight of the Germans although I did receive one hit from a bullet in one wing.'[25]

Nearby, the leader of the 144 Squadron section sighted another two merchant ships and he called for a second attack. But the Beaufighters had lost formation, they had expended all their rockets and the ships were putting up an intense and accurate flak barrage, so the attack was aborted. They wheeled to starboard, toward the east coast of Denmark and towards Langham. Near the coast a 455 Squadron section came across a large a ferry barge that appeared to be carrying troops. The Beaufighters were now down close to the sea. Led by Flying Officer Bob Cristofani, four of them swept across the ferry, firing long bursts of cannon that exploded all over the vessel. The pilots banked away, left and right, and re-set their course for Langham. The end of the War was now very close, and this was the last shipping strike for 455 Squadron.

Next day 455 Squadron ferried their aircraft to nearby North Coates and that afternoon thirteen crews joined another patrol to Kiel Bay. They ran into a wall of black rain clouds over Denmark and the formation leader, Wing Commander Tony Gadd of 144 Squadron RAF, called off the patrol. The aircraft and crews from Dallachy returned to their base in Scotland next afternoon.

Two mornings later, 6 May 1945, seven crews were called on Operations standby. Jack Tucker recalled,

> A big strike was ordered for Bergen, very dangerous. Intelligence showed that German U-boats were making for Bergen. We thought that Germany might be going to continue the war from Norway. At the briefing we were told, 'Gentleman, we expect heavy losses'. I had my foot on the lower rung of the aircraft when transport drove up and said the war was over and the strike was off.[26]

Epilogue

A T last, on 7 May 1945 representatives of Germany's three fighting services congregated at Eisenhower's Headquarters at Rheims. In the presence of senior American, Russian, British and French Officers, the German leaders surrendered to the Western and Russian commands. Jodl and von Friedeberg, acting for Doenitz, signed an instrument which provided for the unconditional surrender of all German forces on all fronts. Having signed, Jodl asked to be able to speak. He said in German, 'With this signature, the German people and the German armed forces are for better or worse delivered into the victor's hands...In this hour I can only express the hope that the victor will treat them with generosity.' There was no reply. Next day the act of surrender was ratified at Russian headquarters in the German capital, Berlin.

There were celebrations at Dallachy. On the evening of the 7th, a bonfire was lit and the men went to an all-ranks dance, with free beer. On 8 May 1945, the 455 Squadron Operations Record Book contains this entry, 'Today is Victory Day in Europe. The Station Commander authorised a stand down of 48 hours for the whole station, a skeleton staff only required to remain on duty.' A church service was conducted and the wireless was connected to the Tannoy broadcast system and kept on all day. At the bottom of a printed Mess menu were the words, 'And now for Japan'.

Despite the relief and the jubilation, Coastal's Air Marshal Sholto Douglas had no intention of letting his Command relax. He had sent this signal to all his groups and stations several days earlier, marked Immediate and Personal for commanders,

> In spite of surrender of German forces on the Continent there is as yet no indication that they contemplate surrender in Norway. We may, therefore, expect the continuance of intense U-boat operations from Norwegian bases. All ranks must realize that for Coastal Command the war goes on as before. We started first we finish last. I call upon all squadrons for a great final effort against an old enemy. It falls to Coastal Command to strike the final blow against the enemy's one remaining weapon.[1]

So, on 11 May, 455 Squadron was back training. They practised formation flying and rocket attacks. They made many escorting sorties over HMS *Devonshire* which carried home the Norwegian Crown Prince Olaf. They escorted Prince Olaf again on 12 May, continued training for several more days, and flew air-sea rescue patrols. John Ayliffe says,

> It was a rather turbulent time because it was felt that some of the enemy forces in Norway may not surrender and there was the possibility of leading Nazi figures escaping from that country. In fact I had to persuade one enemy vessel to heave to by firing across its bow and to direct one of the naval escorts to investigate it.[2]

'A splendid example of devotion to duty': Pilot Officer John Ayliffe, a clerk with the South Australian Railways, flew with 455 Squadron from March 1943 until May 1945, completing a full operational tour. He was awarded the Distinguished Flying Cross in June 1945 for his determination in pressing home numerous attacks on enemy shipping.

There was another rumour. Jack Cox says, 'After VE Day everyone expected to convert to Bristol's Beau replacement (the Brigand) and fly to the Far East as a Squadron'.[3]

On 20 May 1945, Wing Commander Colin Milson was called to RAAF Headquarters in London and was told that the Squadron was to be disbanded. Even so, on the following day, six new crews arrived in the Squadron on posting from operational training units. On that day, 21 May 1945, in the Squadron's last operation of the War, Pilot Officer Peter Ilbery and Warrant Officer Ian Murray flew an anti U-boat patrol from Lindesnes to Utvaer Light on the Norwegian coast. They flew as far north as Bergen, found nothing and returned to Dallachy. On 22 May 1945, all 455's aircraft were grounded pending further instructions for the disbandment of the Squadron.

The Squadron received a signal from Headquarters 18 Group which stated that from Friday 25 May 1945, all 455 Squadron aircraft were to be withdrawn from the line together with aircraft from 144 Squadron RAF and 404 Squadron RCAF. All aircrew were to be held at Dallachy pending disposal while all ground staff were to be posted, mainly to other RAAF Squadrons and servicing echelons. At the end, the Squadron had on charge twenty-one Beaufighters, and there were seventy aircrew and nearly one hundred and twenty ground staff on strength.

A number of new awards were announced. Wing Commander Colin Milson received a Bar to his DSO and both Tommy Thompson and Ralph Jones RAF (Milson's navigator) received Bars to their DFCs. DFCs were awarded to Neil Smith, Frank Proctor, Mick Jackson, Bob McColl, Jack Cox, Ivor Gordon, Noel Turner and John Ayliffe. On that same day, the Commanding Officer of 455 Squadron, Wing Commander Colin Milson, issued a message to the Squadron. He wrote,

> On the eve of the disbandment of the Squadron, I wish to convey my most sincere thanks and gratitude to all members of the Squadron for their willing co-operation and support during my time as Commanding Officer. The results of the Squadron have been a credit to everyone of you and I know that we will all feel very proud, in time to come, to be able to say 'I was with No. 455 Squadron'. In conclusion I wish you all good luck and every success in the future.[4]

When 455 Squadron operations finally ended at Dallachy, there were only eleven aircrew still with the Squadron from those that had been with the Squadron when it arrived at Langham just over twelve months earlier. Of these, there were only two original Langham crews; Neil Smith and his navigator Forbes Macintyre, and John Ayliffe with his navigator Albert Vigor. The Squadron's records list its progressive total of enemy shipping sunk: Ten merchant vessels (30,500 tons), one U-boat, four M-Class minesweepers and three escort vessels.

* * *

In her most comprehensive study of Coastal Command's anti-shipping campaign during the Second World War, Doctor Christina Goulter has made this assessment of the efficacy of that campaign,

> What certainly helped in a direct sense to give the Allies a military advantage during the last part of the War was the way in which Coastal Command's anti-shipping campaign (more so than the passive minelaying effort) diverted from the main operational theatres German manpower and war materials to defend shipping...By having to replace sunken vessels and repair those damaged, the Germans were also losing construction materials which could have been utilized in naval vessels, and the shortage of quality steel plate throughout the War, particularly during 1944, was a major constraint on U-boat construction.
>
> So, although the cost of mounting an anti-shipping offensive was heavy, it should be viewed as an important complementary action, both to strategic bombing and to the campaign against the U-boat menace in the Atlantic.[5]

* * *

At the time of Germany's surrender there were over ten thousand RAAF personnel in Britain with only a quarter of them in RAAF units. The remainder were spread amongst operational RAF squadrons, training units and in miscellaneous RAF units. The problem of quickly unscrambling such a large proportion of Australians from so many different units was made more difficult by an acute shortage of accommodation at the personnel reception centre at Brighton. The general pattern of repatriation was that recovered prisoners of war, if medically fit, should receive first preference. Then, Australians from RAF operational and training units surplus to requirements should return to Australia for possible further employment against Japan. As Empire Air Training Scheme units disbanded their members were also added to the stream for repatriation. During May and June 1945, aircrew holding units were formed and RAAF personnel were sent to these units in Nottinghamshire, Suffolk and Cumberland. Large repatriation drafts occurred in June, August, September and October. The *Aquitania* had over two thousand Australian aircrew on board when it left Britain for Australia in October 1945. Warrant Officer Ivor Gordon recalled that they had to remain at Dallachy for several weeks after VE Day. He says,

> The Air Officer Commanding Coastal Command broadcast a message saying that it was anticipated that the Germans in Norway might fight on and we may be needed for further operations. Later on I met Colin Milson in a cafe in London. He told us that the Americans had dropped an atomic bomb in Japan and that the war in the Pacific was over.
>
> Once we left Scotland we used to hang around the 'Boomerang Club' at Australia House in London. At one stage the word was out that there was work at a film studio in Durham. A lot of us were in the film as extras and we were paid three pounds per day for that work.
>
> The reception centre at Brighton filled up and we formed a number of holding units. We had to report in every three weeks to see when we would be going home. I left on the liner *Aquitania*. When it arrived at Fremantle in Western Australia it was too big to get into the harbour and we were brought to the wharf in lighters. I managed to be on the very last boat, to the annoyance of my fiancee.[6]

✳ ✳ ✳

Late in 1946, a group of former 455 Squadron members began meeting once a month at the RAAF Association rooms in Sydney. By the next year they were granted the status of a Branch with the NSW Division of the Air Force Association. Jack McKnight says,

> Jack Davenport was our first president and is still in the chair. We continued meeting on the first Friday of each month and could count on 20 to 30 turning up. Members started to dwindle over the years and the big turning point was the introduction of random breath testing for drivers in NSW. Attendance dropped rapidly to about half a dozen regulars and finally some of these succumbed to illness and changes of address and meetings as such ceased. For some years now our only get together has been on Anzac Day and one or two special occasions. Although the contact may be infrequent, the bonds are sincere.[7]

In January 1991, Peter Ilbery wrote to the Sheriff of Morayshire in Scotland,

> It remains a compelling idea to me that there should be a mark of respect in the vicinity of the [Dallachy] base from which they [the aircrew lost in operations from RAF Dallachy] made their last flights...The worst losses from RAF Dallachy were on Black Friday, 9th February 1945, when nine Beaufighters of the thirty-two strong Strike Force sent against shipping in Forde Fiord failed to return...If there is indeed no memorial [at Dallachy], with your encouragement, I would take the initiative in pressing for one.[8]

By coincidence, the week that the Sheriff of Morayshire received Peter Ilbery's letter a local newspaper had printed an article about the launch of a two thousand pound fund for just such a memorial.

'History flew beside them': The Memorial to the Dallachy Strike Wing was unveiled at Dallachy on 30 July 1992. It stands at the entrance to the Dallachy station, now farmland, although the control tower and runways can still be seen.

On July 30 1992, Peter Ilbery was back at the Dallachy Field in Scotland. A thousand people had gathered near the former main gate of the airfield. Peter Ilbery unveiled a memorial, a stone cairn in the shape of an aircraft's tail fin, containing stones from each of the countries that had been represented at Dallachy. Peter Ilbery said during his address, 'The memory of the happenings here may have faded but not the emotions aroused by those events. It is those powerful emotions that have brought us together.'[9] At 2.30 pm there was a fly-past by an RAF Nimrod and two RAF Buccaneer aircraft. Next day, Peter Ilbery drove along the newly named 'Beaufighter Road' which links Upper Dallachy to Nether Dallachy. He recalled,

One turns onto it from the Buckie Road and dropping down over the ridge it goes into a wood of pines which conceals the old perimeter track about the previous 455 dispersals. Probably my clearest memory of Dallachy is taxiing round this perimeter to runway 23 for formation take-offs into the south towards the gap in the hills (with the Sergeants' Mess to port). One vehicle driven by a 144 crew with the navigator at the wheel and the pilot in the passenger seat picked its way a little gingerly but nevertheless ceremoniously over this crumbling concrete main runway. Beaufighter Road then continues further around the perimeter to the outskirts of the village. The perimeter track, navigable with care, then brought us back to the control tower which has largely withstood the elements.

Then up the ridge to Ops; the farm house and buildings adjacent flow over it and with a wall missing and a good deal of damage it has become a vehicle storage utility. The eastern wall of the old Ops main chamber still shows on its cracking plaster the following tabulated headings:

Captain   Aircraft   ETD   ETA   Remarks [10]

Now, this memorial at Dallachy provides a local focal point for the observance of Remembrance Day, a Memorial to those without known graves. 455 Squadron men whose last flights were from Dallachy are Flight Sergeant R.E. Bull, Flying Officer W.J. Edwards, Warrant Officer W.T. Furlong, Flying Officer J.A. Hakewell, Warrant Officer K.L. Hamilton, Flying Officer G.W. Hassell, Flying Officer G.H. Hammond, Flight Sergeant G.A. Henry, Flight Sergeant I.H. Jury, Flying Officer E.J. Loonam, Warrant Officer W.D. Mitchell, Flight Sergeant F.G. Sides and Pilot Officer A.E. Winter.

The sacrifices of the men who flew their last flight from Langham are also remembered. On Sunday 3rd October 1993, a plaque was unveiled at a church at Langham dedicated to 455 Squadron RAAF and 489 Squadron RNZAF. It carries this inscription,

> From the fields between this church and the sea during the wartime summer of 1944, a small band of young men flew in defence of these islands. Most of these ardent volunteers had journeyed across half the globe to our aid in a time of desperate need. The many successful attacks on enemy shipping made by the pilots of 455 Squadron RAAF and 489 Squadron RNZAF Torpedo Bomber Squadrons and their British navigators made a valuable contribution to the preservation of our freedom.
>
> Sadly, many of these young men were destined never to return home again, but to be lost somewhere across a waste of seas, with their final resting place remaining unknown to this day. May their sacrifice for us never be forgotten.

The airfield at Langham can still be recognised, although one of the runways is occupied by several long huts – a turkey farm. The land is covered in fields of barley. The control tower is easily recognisable, with its distinctive square shape and flat roof, now used as an office.

* * *

The Commanding Officer from 455 Squadron's Langham days, Wing Commander Jack Davenport, remained at Headquarters 18 Group until the end of the War. He was one of three brothers who served with distinction with the RAAF in the European theatre. Squadron Leader Philip Davenport completed his first operational tour with 461 Squadron RAAF, flying Sunderland flying boats on anti U-boat operations. His second tour was with 243 Squadron RAF, flying Mosquitoes on anti-shipping strike wing operations from Banff in Scotland. During an attack on Porsgrunn harbour south-west of Oslo (an attack that Jack Davenport had planned), Philip Davenport was shot down and he crash-landed his Mosquito on a small lake, solid with ice. He became a prisoner of war but escaped a few days before the War in Europe ended. Jack Davenport's youngest brother, Pilot Officer Keith Davenport, also flew Sunderland aircraft with 461 Squadron RAAF, on anti-submarine operations. His aircraft was attacked by a German Ju-88 aircraft and severely damaged, and Keith Davenport had to ditch near the Isles of Scilly on the south-west point of Britain. All his crew were rescued.

When the War ended, Jack Davenport was sent from Headquarters 18 Group to Norway in charge of a team to investigate the effectiveness of air attacks on German sea communications. The team travelled in an air-sea rescue boat and actually liberated many of the ports on southern Norway. Later, Jack Davenport did a similar investigation through northern France, Holland, Germany and Denmark. On his return from Europe he was posted as Commanding Officer of RAF Station Banff in Scotland, his command including the station at Peterhead. Then he spent some time at Kodak House in London as Senior Air Staff Officer to look after affairs of the RAAF and Australian airmen in Britain.

Jack Davenport was greatly respected by his men. He had shown great skill at flying, and courage in the face of the enemy and in rescuing his men from crashed and burning aircraft. He had their admiration for having completed two operational tours and they enjoyed his enthusiasm and leadership. The men of 455 Squadron were the best judges and they respected leadership. One veteran of thirty-seven operational sorties said,

> Davenport must undoubtedly have been one of the really outstanding squadron commanders of the air war. To try to analyse the complex compound of talents and virtues which make up his leadership qualities might only cloud that assessment; qualities that he carried through with such distinction in civilian life.

Jack Davenport makes these comments about leadership,

> Know the task, set clear objectives, ensure they are understood. Involve the people, seek their ideas, challenge their abilities. Leadership is obtaining results through the efforts of other people; it requires the challenging in one form or another of every set of information and every situation. Leaders develop better answers and better results than other people.

Jack Davenport is one of the most qualified when it comes to comment about the work of the strike wings. He says, 'The loss of so many aircraft of the strike wings was testimony to the difficulties of their attacks, increased by the rugged terrain of Norway, but also to the aggressive courage and skilful attacks of the aircrew'.

Jack Davenport was back in Australia in March 1946, when he joined Monier Ltd. He was a Director of a number of Australian companies and organisations including the Reserve Bank of Australia, QANTAS Australia, MIM and Alcoa, and he was a Councillor of the Australian War Memorial and the Australian Science and Technology Council. He was made an Officer of the Order of Australia in 1983 and received Australia's highest award, the Companion on the Order of Australia in 1991 for services to Australian Industry and the community. Jack Davenport's navigator from the Beaufighter period, Ralph Jones DFC, RAF, returned to his job with the British Customs service. He has visited Australia to see his many friends from 455 Squadron.

Colin Milson, the 455 Squadron Commanding Officer at Dallachy, was sent from Dallachy to command the Aircrew Holding Unit at Beccles. He also spent some time test-flying the new Bristol Brigand, the successor to the Beaufighter. Early in 1946 Colin Milson returned to Australia with the Beaufighter control column from his favourite Beaufighter S-Sugar . This aircraft had been allocated to 455 Squadron when it first re-equipped with Beaufighters, and was one of the oldest in the Squadron by the end of the War (three times it returned safely from operations on one engine, twice when Colin Milson was flying). In Australia, Colin joined the Brisbane-based Thomas Borthwick meat company as a cattle buyer. Although the Milson family property in Queensland was sold in 1973, he continued to run three properties as Director and Chief Executive Officer of the Sydney company that owned them. Colin Milson died of a heart attack in hospital in Sydney on 14 July 1975, aged 56. His qualities as a brave, professional and skilful leader, airman and leader were widely known. He has been described as having had 'a lot of dash and flair'. Another Squadron pilot says, 'He was interested in his men; a very capable leader mainly because he didn't 'frig about' before attacks'.

Grant Lindeman refused a permanent commission with the RAF, was 'demobbed', spent a year in a London Engineering office, and under the terms of his original Short Service Commission he was paid a five hundred pound gratuity and repatriated on the *Orion* with his wife and small son. He became an oyster farmer at Port Stephens for twenty years, then he took over a cattle property in the Armidale district of New South Wales. Over fifty years later, Grant Lindeman says, 'The character of a Squadron derives from the characters of the people in it. 455 was blessed with a number of colourful characters who all had a well developed sense of humour'.

Lloyd Wiggins left 455 Squadron shortly after it arrived at Dallachy, to be Chief Instructor at an Operational Training Unit. When the War ended he was sent as Commanding Officer at 11 PDRC in Brighton, to assist the return home of Australian airmen. He finally returned to Adelaide as a Valuer and Auctioneer. He has been involved with Legacy and the United Service

Institute in Adelaide for over forty years and he now lives at Aldgate in South Australia with his wife Thelma, who had been the Officer in charge of the WAAF at Langham.

Neil Smith was sent from Dallachy to London after disbandment of the Squadron, to write a report with Squadron Leader John Pilcher on the use of rocket projectiles. He returned home in December 1945 and was discharged soon after. He now lives on a property he calls 'Del-a-Chi' at Narracoorte, South Australia.

Bill Herbert completed thirty-seven operations with 455 Squadron and was three quarters of the way through his operational tour when the War ended. He returned home to university then joined the Royal Australian Navy as a pilot in 1949. Bill Herbert now lives in Canberra. His navigator, Jack Tucker, returned to study Law, then became a leader in his field as a company secretary. He now lives in Melbourne.

Clive Thompson completed two operational tours in Europe. He returned home to South Australia. His navigator, Ivor Gordon, returned to Perth and began a successful business as a licensed surveyor.

One of the Squadron Outriders, 'Flak Jack' Cox finished his first operational tour as the War ended. He was sent back to 11 PDRC to wait for transport home. Jack Cox now lives in Gosford, New South Wales. His navigator, Allan Ibbotson, returned to study mining and became a surveyor on the Great Boulder Mine. He worked as a mining inspector, then finished his career as a Regional Mining Engineer in Karratha. He retired on his 60th birthday, in 1982. He now lives in Rockingham.

John Ayliffe and his navigator, Albert Vigor, completed their operational tours with 455 Squadron. John Ayliffe had joined 455 Squadron in March 1943 and he was one of the Squadron's longest serving Beaufighter pilots. He returned to work for the South Australian Railways and now lives in Adelaide. Albert Vigor married a WAAF in Britain and they travelled to Australia where Albert took up his trade as a butcher in Cairns. Albert died of leukemia in the seventies.

Peter Ilbery returned to Australia to study medicine and several years after qualifying he joined the Air Force Reserve as a radiologist. After postgraduate work at the Atomic Energy Research Establishment at Harwell in England, he became advisor to the Director of Air Force Health Services on the effects of atomic radiation. He attained the rank of Group Captain. Peter Ilbery lives in Canberra. His navigator, Bill Bawden, returned to his wife and child and his china importing business. Peter Ilbery expresses strong admiration for his friend and navigator. Although Bill Bawden was a number of years senior to Peter they had become friends early in their service careers and they both found it very satisfying to form a crew for service with 455 Squadron.

Noel Turner was a Squadron veteran and had finished a full operational tour when the War ended. Back home, Noel and his brother pooled their savings and bought a vehicle service station which they built up over the years to a car dealership. Jack Davenport was a valued customer. Noel agreed with another squadron member who said that 'most of the fellows on the Squadron were pretty good blokes to know, not for their drinking ability but because they never chickened out in a pinch'. Noel retired to Bateau Bay in New South Wales and died in 1994. He is remembered fondly by all who served with him as a courageous and skilful pilot, and a very good friend. He was extremely helpful in providing details and valuable insights during the preparation of this Squadron story, treating his own important role with great humility. Noel Turner flew with many navigators, and the last was Bert Iggulden, known as 'Imperturbable Bert' by some. An RAF navigator, Bert Iggulden said of the Australians, 'I enjoyed being with them. The morale was the best. I always regarded 455 as a top squadron, with 489.' Bert Iggulden summed up his thoughts this way, 'You'd need to be insane to enjoy this pastime, I believe however that the strikes did much to shorten the war.'

Bob McColl had been on his 51st operational sortie when he was shot down over Forde Fiord on 9 February 1945. He was taken prisoner and returned to Britain after release from a prisoner of war camp in May 1945. He returned to his farm in Koorawatha, New South Wales. His wife Katie, who had been a WAAF at Langham joined him some nine months later. Bob died suddenly from a heart attack on 2 December 1970. Katie later returned to England, where she now lives in Norfolk, near Langham.

Vic Pearson returned home, then decided to stay in the Air Force and spent much of his time in 87 Squadron, engaged in survey operations. He was awarded a Commendation for Meritorious Service in the Air for that work. His last posting was as Administration Officer of Number 1 Squadron at Amberley in Queensland, which was re-equipping with the F-111 bomber. He now lives in Brisbane.

David Whishaw did a course in Agriculture in New Zealand, and then returned to a mixed farm in Tasmania. David says that 'I look back on that tour of operations with some satisfaction but also of opportunities not always grasped. As a 21 year old I had no further ambition at that time than just to be a pilot in action'.

Wally Kimpton returned home to his wife and his job in Perth with Bankers and Traders Insurance. He retired as state manager in 1974. Wally Kimpton has commented on the damage that his Beaufighter sustained during the strike on 6 July 1944 – 'Not before or since had my aircraft been damaged in any way and so I eventually finished my tour on a fortunate note convinced that luck plays a most important part in our lives.' He also says, 'I should emphasize the role played by our navigators. They were always at the mercy of their pilots and they must have had nerves of steel to carry on week after week so cheerfully.'

Ted Watson was sent to America for a Rocket Projectile course in April 1945. He returned to Australia after three years away and was posted to Victoria Barracks in Melbourne. Ted was discharged from the Air Force in September 1945 and went back to his job as a teacher in Sydney. He was later appointed as an Inspector of Schools in New South Wales. Ted had a heart problem and on 15 November 1973 he was driving to visit a high school near Lake Macquarie. He drove to a nearby surgery, walked in and asked a waiting patient to lock his car. He then had a massive heart attack.

Keith Carmody became a cricket personality after the War and has been credited with having Western Australia admitted into the Sheffield Shield cricket competition. He played for Australia and was Captain of the Western Australian cricket team for a number of years.

Harold Spink became the sales manager for an international packaging company and served on the State Council of the Air Force Association for some years. Harold died in Sydney in July 1991, several years after a heart by-pass operation.

Ian Murray bought a grazing property in Queensland and remained there for thirty-four years before he retired to Noosa, where he continued to work as a photo journalist.

Steve Sykes, a grazier from Goulburn, was killed in an aircraft accident in Australia in 1957.

Jack MacDonald returned to Australia to work in local and export marketing, a part-time university commerce course, and as a company General Manager. He later served in Singapore and Japan as a Government Trade Commissioner. After retirement, Jack studied Chinese and escorted tours to China. He now lives in East Melbourne.

Bill Waldock returned to his job as an accountant with a mines management company in Perth, and later ran a general store. When the Korean War began he rejoined the RAAF and was posted to Japan as a fighter controller and operations officer. Bill also served with 77 Squadron in Seoul. He returned to Australia and served with the RAAF in Pearce and Woomera, and at Air Headquarters in Melbourne and Canberra. He commanded Number 3 Control and Reporting Unit at Williamtown and Number 114 Mobile Control and Reporting Unit at Amberley. Bill retired from the RAAF in 1973. He lives in Perth.

For the men and women of 455 Squadron, the Strike Wing Memorial at the Dallachy field contains these lines from the English poet C. Day Lewis,

> History flew beside them, and bright fame
> Arches her wings above the cloudy wars

Appendix 1

455 Squadron Operations Summary January 1944 to May 1945

This Appendix summarizes the major operations and events for 455 Squadron over the period of its operations with Beaufighters

Date	Operation	Result

RAF Leuchars

Date	Operation	Result
10 Dec 43	Last 455 Sqn operation with Hampdens	Rover patrol off the southern Norwegian coast by three 455 Sqn aircraft. Nothing sighted
13 Dec 43	455 Sqn withdrawn from operations for re-equipping with Beaufighters	
29 Dec 43	Beaufighter flying training commences	Five 455 Sqn pilots went solo. One 455 Beaufighter crashed and burst into flames during a demonstration flight. Both crew killed (F/O P.S. Gumbrell RAF and F/O Toombs 489 Sqn RNZAF in Beaufighter LZ 193)
29 Dec 44	The last 455 Sqn Hampden aircraft left RAF Leuchars for disposal	
23 Feb 44	455 Sqn completed Beaufighter conversion training	
1 Mar 44	455 Sqn operational with Beaufighters as part of the 'Leuchars Wing' with 489 Sqn RNZAF	
6 Mar 44	Convoy strike off the southern Norwegian coast by the Leuchars Wing including eight 455 Sqn Beaufighters	Seven ships damaged. The 994 ton German merchant vessel *Rabe* was sunk
10 Mar 44	Morning patrol along the Norwegian coast in the vicinity of Helliso Light by nine 455 Sqn Beaufighters	Nothing sighted
10 Mar 44	Afternoon anti-shipping patrol by the Leuchars Wing along the Norwegian coast north from Obrestad Light, including four 455 Sqn Beaufighters	Visibility poor, nothing sighted

16 Mar 44	Fifteen crews from 455 Sqn departed for a detachment to RAF Skitten with ten crews from 489 Sqn RNZAF	
22 Mar 44	Patrol along the Norwegian coast from RAF Skitten by nine 455 Sqn Beaufighters accompanied by 489 Sqn RNZAF	No shipping sighted
23 Mar 44	Afternoon patrol along the Norwegian coast from RAF Skitten including nine 455 Sqn Beaufighters	Bad weather encountered, formation broke up
25 Mar 44	Afternoon anti-shipping patrol from RAF Wick by eight 455 Sqn Beaufighters, with three Beaufighters from 489 Sqn RNZAF	Three armed trawlers sighted. Weather unsuitable for attack
26 Mar 44	Notification received that 455 Sqn and 489 Sqn were to be transferred to RAF Station Langham in Norfolk, by 15 Apr 1944	
28 Mar 44	Nine 455 Sqn Beaufighters took part in an anti-shipping patrol off the Norwegian coast with five Beaufighters from 489 Sqn RNZAF	No enemy shipping sighted and the aircraft returned to RAF Skitten
28 Mar 44	Crews returned to Leuchars after the detachment to RAF Skitten	
2 Apr 44	455 Sqn taken off the fighting strength of 18 Group and transferred to 16 Group Coastal Command	
6 Apr 44	455 Sqn advance party moves to RAF Langham by train	
11 Apr 44	Sqn main party departs RAF Leuchars by train	
14 Apr 44	455 Sqn ordered to commence training for a bombing role against German shipping	

RAF Langham

15 Apr 44	455 Sqn move to RAF Langham complete	
19 Apr 44	Morning and evening reconnaissance patrols along the Dutch coast by eight 455 Sqn Beaufighters operating independently	Convoys sighted. Light flak from ships and shore batteries. Sighting reports sent
21 Apr 44	Four 455 Sqn Beaufighters conducted independent reconnaissance patrols along the Dutch coast	One crew sighted shipping near Den Helder Harbour. Sighting report sent to base
22 Apr 44	Reconnaissance patrol along the Dutch coast by the Langham Wing including eight 455 Sqn Beaufighters. Led by W/Cdr J.N. Davenport escorted by Spitfires from 611 Sqn RAF	Nothing sighted

24 Apr 44	First-light reconnaissance in force by the Langham Wing including six 455 Sqn Beaufighters	Sighted an 800 ton trawler
25 Apr 44	Four 455 Sqn crews conducted independent first-light reconnaissance patrols along the Dutch coast	No enemy shipping sighted
26 Apr 44	Two 455 Sqn crews accompanied ten Beaufighters from 489 Sqn RNZAF on a patrol along the Dutch coast, escorted by Spitfires from 611 Sqn RAF	No enemy shipping sighted
27 Apr 44	Reconnaissance in force along the Dutch coast by the Langham Wing including six 455 Sqn Beaufighters	Two convoys sighted
30 Apr 44	Four 455 Sqn Beaufighters on separate last-light reconnaissance patrols	Two convoys sighted
1 May 44	Aircraft crashed on take-off	One 455 Sqn crew member killed (P/O N.A. Wilson in UB-W, NE 353)
4 May 44	Early morning reconnaissance to the English Channel by the Langham Wing, including sixteen 455 Sqn aircraft. Led by W/Cdr J.N. Davenport	Attack on two Royal Navy motor torpedo boats
6 May 44	Convoy strike off the Dutch coast by the Langham Wing including twelve 455 Sqn Beaufighters	One merchant vessel torpedoed. Many other vessels damaged. One 455 Sqn crew lost (F/Lt B. Atkinson RAF and F/Sgt J. A. Whitburn in UB-M, NE 206)
8 May 44	Six 455 Sqn Beaufighters took off in pairs for reconnaissance patrols along the Dutch coast	Several convoys sighted and reports sent to base
9 May 44	Armed reconnaissance along the Dutch coast by the Langham Wing including ten 455 Sqn Beaufighters	No enemy shipping sighted
11 May 44	Independent shipping reconnaissance patrols off the Dutch coast by six 455 Sqn Beaufighters	No shipping sighted
14 May 44	Convoy strike off the Dutch coast in the vicinity of Ameland by the Langham Wing, including twelve 455 Sqn Beaufighters. Led by W/Cdr J.N. Davenport with a Mustang fighter escort from Coltishall	Dutch vessel *Vesta* (1854 tons) and German minesweeper M.435 (637 tons) sunk
17 May 44	Four independent shipping patrols along the Dutch coast	Convoy sighted, report sent
19 May 44	First-light reconnaissance patrol along the Dutch coast by four 455 Sqn Beaufighters	Two large convoys sighted
20 May 44	Last-light anti E-boat patrol along the Dutch coast from The Hague to Den Helder by twelve 455 Sqn Beaufighters	No sightings

22 May 44	First-light anti E-boat patrol by the Langham Wing along the French coast, including six 455 Squadron Beaufighters	No sightings
24 May 44	Twelve 455 Sqn crews on a first-light anti E-boat patrol along the French coast	Patrol recalled
1 Jun 44	First-light anti E-boat patrol by the Langham Wing along the Dutch coast from Ijmuiden to Terschelling including six 455 Sqn aircraft, with a Mustang fighter escort	Nothing sighted. First operation where 455 Sqn carried 250 lb and 500 lb bombs
2 Jun 44	First-light anti E-boat patrol off the French coast from Dieppe to Cap Balfour by the Langham Wing, including six 455 Sqn Beaufighters with a Spitfire escort	No enemy shipping sighted
3 Jun 44	First-light anti E-boat patrol off the French coast by the Langham Wing including six 455 Sqn Beaufighters, escorted by a sqn of Spitfires	No enemy shipping sighted
5 Jun 44	Last-light anti E-boat patrol along the French coast by the Langham Wing including five 455 Sqn Beaufighters	No result
6 Jun 44	Evening patrol along the French coast from Boulogne to Fecamp by five 455 Sqn Beaufighters, in company with three 489 Sqn RNZAF Beaufighters. The patrol departed from RAF Manston in Kent	E-boats sighted by 489 Sqn but no attack made due to failing light
7 Jun 44	First-light E-boat attack in the vicinity of Cherbourg by the Langham Wing, including six 455 Sqn Beaufighters	One E-boat damaged
8 Jun 44	Afternoon reconnaissance patrol along the Dutch coast, from Terschelling to Heligoland by two 455 Sqn Beaufighters	One Beaufighter crashed into the sea with engine trouble. The crew was lost (F/O F.O. Williams and F/Sgt W.A. Roach in UB-A, NE 200)
8 Jun 44	Last-light anti E-boat patrol along the Dutch coast from Ostend to Ijmuiden by six 455 Sqn Beaufighters	No enemy shipping sighted
8 Jun 44	Last-light shipping reconnaissance along the Dutch coast convoy routes from Terschelling to Heligoland, by two 455 Sqn Beaufighters	Two large convoys sighted and reports sent back to base
10 Jun 44	First-light anti E-boat patrol along the Cherbourg Peninsula by six 455 Sqn Beaufighters. Led by S/Ldr A.L. Wiggins	No enemy shipping sighted. Aborted attack made on a friendly vessel in the Free Bombing Area
10 Jun 44	Evening reconnaissance patrol by two 455 Sqn Beaufighters	Convoy of six merchant vessels with escorts sighted. Patrol fired on by accurate flak. Sighting report sent to base
11 Jun 44	Anti E-boat patrol along the Dutch and Belgian coasts by six 455 Sqn Beaufighters.	455 Sqn navigator bailed out from a damaged aircraft, which returned to Langham safely.

Jun 44	Early morning patrol and strike off the Hook of Holland by twelve 455 Sqn Beaufighters, in company with Beaufighters from the North Coates Wing	Five ships severely damaged. One 455 Sqn crew shot down and captured (F/O D.K Carmody and F/O G.C. Docking in UB-Z, NE 668)
14 Jun 44	Early morning anti E-boat patrol off the Dutch coast by the Langham Wing, including six 455 Sqn Beaufighters. Four minesweepers attacked	No results observed
15 Jun 44	Convoy strike in the vicinity of Ameland by the Langham Wing including eleven 455 Sqn Beaufighters. Led by W/Cdr Gadd 144 Sqn RAF and escorted by ten Mustangs of 316 (Polish) Sqn from Coltishall	Dutch experimental vessel *Amerskerk* (7900 tons) sunk. Belgian E-boat depot ship *Gustav Nactigall* (3500 tons.) sunk. German Minesweeper M.103 (772 tons) sunk
16 Jun 44	Anti E-boat patrol off Ostend by six 455 Sqn Beaufighters	No result observed
19 Jun 44	First-light reconnaissance in force to the vicinity of Borkum by the Langham Wing with six Beaufighters from 236 Sqn RAF (North Coates), including thirteen 455 Sqn Beaufighters	No enemy shipping sighted
24 Jun 44	First-light anti E-boat patrol from Gravelines to the Hook of Holland by the Langham Wing including three 455 Sqn Beaufighters	Six unidentified vessels seen inside the Hook of Holland
25 Jun 44	First-light anti E-boat patrol from Gravelines to the Hook of Holland by the Langham Wing including six 455 Sqn Beaufighters	No sighting
27 Jun 44	A strike force of twenty-nine aircraft from the Langham Wing , including fifteen Beaufighters from 455 Sqn set course for the Dutch coast	The patrol was aborted because of bad weather
28 Jun 44	Early morning reconnaissance by two 455 Sqn Beaufighters.	Convoy sighted off Heligoland
28 Jun 44	Anti E-boat patrol from Dunkirk to the Hook of Holland, by six 455 Sqn Beaufighters. A convoy of minesweepers was attacked	No result confirmed
29 Jun 44	Early morning anti E-boat patrol from Dunkirk to the Hook of Holland by the Langham Wing including six 455 Sqn Beaufighters	A convoy was attacked off the Hook of Holland. Many vessels damaged. One 455 Sqn Beaufighter shot down and one crew member killed (F/O E.F. Collaery in UB-H, LZ 192)
29 Jun 44	Evening anti E-boat patrol from Gravelines to the Hook of Holland by twelve 455 Sqn Beaufighters	Very poor visibility, no sightings
1 Jul 44	Training accident. A 455 Sqn Beaufighter crashed during formation flying practice.	One 455 Sqn crew killed. (F/O J.U. Billing and F/O T.O. Edwards in UB-C, NE 773).

Jul 44	First-light anti E-boat patrol from Gravelines to the Hook of Holland by the Langham Wing, including six 455 Sqn Beaufighters. Led by S/Ldr C.G. Milson	Nothing sighted, bad visibility
Jul 44	Last-light anti E-boat patrol from Gravelines to the Hook of Holland by six 455 Sqn Beaufighters, led by S/Ldr A.L. Wiggins	Six small vessels attacked near Calais, no result observed. Intense flak encountered
Jul 44	First-light anti E-boat patrol from Dunkirk to Ostend by the Langham Wing including six 455 Sqn Beaufighters	Patrol returned due to poor weather
Jul 44	Morning armed reconnaissance patrol to the vicinity of Heligoland by the Langham and North Coates wings, including eleven 455 Sqn Beaufighters	Heavy low cloud prevented an attack
ᴵ Jul 44	Evening attack on a convoy of ten merchant vessels off the Dutch Coast by the Langham and Strubby wings, including ten 455 Sqn Beaufighters	German merchant vessels damaged and sunk. Three 455 Sqn crew members killed. (F/O W.M. Barbour and F/O F.G. Dodd RAF in UB-F, NE 348, and F/Sgt J. Costello in UB-U, LZ 194). One 455 Squadron crew member captured (F/Sgt R. Taylor)
ᵃ Jul 44	First-light anti E-boat patrol from Dunkirk to the Hook of Holland by the Langham Wing, including four 455 Sqn Beaufighters	Nineteen unidentified vessels sighted
ᵃ Jul 44	Reconnaissance patrol off the Frisian Islands by a single 455 Sqn Beaufighter (F/O D. Whishaw and F/O J. Belfield RAF)	One German R-boat attacked and damaged
ᵃ Jul 44	Evening anti E-boat patrol from the Hook of Holland to Dunkirk by six 455 Squadron Beaufighters, led by F/O E.W. Watson	Crews saw what appeared to be a naval engagement
10 Jul 44	Afternoon reconnaissance patrol to the Dutch coast by two 455 Sqn Beaufighters	One 455 Sqn Beaufighter crashed into the sea. Both crew members killed (F/Sgt M.M. Roberts and F/Sgt J.T. Andrew in UB-A, NE 760)
13 Jul 44	Reconnaissance patrol along the Frisian Islands by a single 455 Sqn Beaufighter (F/Sgt R. Walker RAF and F/Sgt T. Rabbitts RAF)	A convoy of twelve merchant vessels and escorts were seen in the mouth of the Weser Estuary. A sighting report was sent
15 Jul 44	Convoy strike off Mandal by the Langham and Strubby wings, including twelve 455 Sqn Beaufighters. Led by W/Cdr J.N. Davenport with a Mustang fighter escort	Four merchant ships and many escorts damaged
17 Jul 44	Evening reconnaissance to the Frisian Islands by a single 455 Sqn Beaufighter	A convoy of six merchant ships was sighted

(P/O A.E. Jones RAF and F/Sgt
W.H.H. Iggulden RAF).

18 Jul 44	First-light anti E-boat patrol from Gravelines to the Hook of Holland by six 455 Sqn Beaufighters.	Nothing sighted
18 Jul 44	Evening reconnaissance to the Frisian Islands by a single 455 Sqn Beaufighter (W/O J.A. McLean and F/O H.J. Wilding RAF). Convoy encountered in the vicinity of Nordernay which opened fire on the Beaufighter.	Sighting report sent
18 Jul 44	Evening anti E-boat patrol from Gravelines to the Hook of Holland by six 455 Sqn Beaufighters. Led by F/O R.C. McColl.	Accurate heavy flak encountered. No shipping sighted. Three flying bombs were seen on the return journey
20 Jul 44	Afternoon armed reconnaissance patrol along the Norwegian coast from Lister by the Langham and North Coates wings, including eleven 455 Sqn Beaufighters.	Intense heavy flak encountered from Lister. The formation broke up and returned to base
21 Jul 44	Evening convoy strike off Heligoland by the Langham and Strubby wings, including eleven 455 Sqn Beaufighters. Led by S/Ldr C.G. Milson.	Finnish merchant vessel sunk. German minesweeper sunk
23 Jul 44	Reconnaissance in force by the Langham and North Coates wings to the south Norwegian coast including ten 455 Sqn Beaufighters. Led by S/Ldr A.L. Wiggins with a Mustang fighter escort.	Fishing vessels sighted. Bad weather encountered
25 Jul 44	First-light anti E-boat patrol from Gravelines to the Hook of Holland by six 455 Sqn Beaufighters.	Heavy flak encountered. Several minesweepers seen. Formation broke up and returned to base
26 Jul 44	Last-light armed reconnaissance in force to the Den Helder area by five 455 Sqn Beaufighters, searching for an Elbing Class destroyer.	Patrol height 6000 ft. No sighting
27 Jul 4	Evening reconnaissance to the Frisian Islands by a single 455 Sqn Beaufighter.	Two merchant vessels sighted near Wangerooge. Heavy flak from Heligoland. Sighting report sent
28 Jul 44	Anti E-boat patrol from the Hook of Holland to Den Helder by six 455 Sqn Beaufighters.	Nothing sighted
30 Jul 44	Evening armed reconnaissance patrol from Lister to Stavanger by the Langham Wing with Beaufighters from 143 Sqn RAF (North Coates), including twelve 455 Sqn Beaufighters. Led by W/Cdr J.N. Davenport with a Mustang fighter escort.	Nothing sighted on the patrol. Attacked by German fighter aircraft as they returned. The Mustang escort attacked the German fighters, shooting several down.

4 Aug 44	Evening reconnaissance patrol off the Dutch coast by a single 455 Sqn Beaufighter (W/O J.C. Ayliffe and W/O A.T. Vigor)	Several small merchant ships reported
4 Aug 44	Night operations by the Langham Wing operating from Thorney Island, against E-boats attacking Allied shipping in the "Overlord" invasion area. Led by W/Cdr J.N. Davenport	Several attacks made. Shipping damaged
8 Aug 44	Afternoon convoy strike off Stavanger by the Langham Wing with Beaufighters from 254 Sqn RAF (North Coates), including ten 455 Sqn Beaufighters. Led by W/Cdr Burns 254 Sqn RAF escorted by 48 Mustang fighters	Norwegian merchant vessel sunk. Several armed trawlers damaged
9 Aug 44	455 Sqn Beaufighter crashed on return to Langham after an air test	One crew member hospitalised (W/O N.P. Turner)
10 Aug 44	Evening strike off Heligoland by twenty-four Beaufighters from the Langham Wing (including twelve 455 Sqn Beaufighters) with fifteen Beaufighters from 254 Sqn RAF (North Coates). Led by F/Lt J.M. Pilcher	German merchant vessel *Santos* (5943 tons) sunk. Three 455 Sqn crews lost (W/O W.T. Jones and F/Sgt H.E. Brock in UB-L, NE 340, F/O L.A. Kempson and F/O R. Curzon RAF in UB-D, KW 277, W/O G.E. Batchelor RAF and P/O H.R. Morris RAF in UB-H, NT 959)
11 Aug 44	Evening reconnaissance off Heligoland by the Langham and North Coates Wings including six 455 Sqn Beaufighters. Led by S/Ldr C.G. Milson with a Mustang fighter escort	Very poor visibility. Nothing sighted
13 Aug 44	Early morning convoy strike near Heligoland by the Langham Wing with Beaufighters from 254 Sqn RAF (North Coates), including ten 455 Sqn Beaufighters. Led by F/Lt Ford 254 Sqn RAF with a Spitfire fighter escort	German Minesweeper M.383 (637 tons) sunk, German Flak ship *Preussen* (425 tons) sunk
16 Aug 44	Early morning reconnaissance off the Dutch coast by the Langham and North Coates Wings, including six 455 Sqn Beaufighters. Escorted by Spitfire fighters	Nothing sighted
24 Aug 44	Morning reconnaissance by the Langham and North Coates wings, including nine Beaufighters from 455 Sqn.	Nil result. One 455 Sqn Beaufighter crashed and caught fire on landing, (UB-O, NE 326). Crew (W/O J.A. McLean and F/Sgt W.A. McHugh) unhurt
25 Aug 44	Combined Langham/North Coates Wing strike near Borkum including fourteen 455 Sqn Beaufighters. Led by W/Cdr J.N. Davenport with a Mustang fighter escort	German minesweeper M.347 (637 tons) sunk
27 Aug 44	Anti E-boat patrol by the Langham Wing including six 455 Sqn aircraft	Nothing sighted

29 Aug 44	Combined Langham/North Coates Wing strike off Heligoland including thirteen 455 Sqn Beaufighters. Led by W/Cdr Burns 254 Sqn RAF with a Mustang fighter escort	German Mine Destructor ship *Hermes-Sperrbrecher 26* (2503 tons) sunk. German Mine Destructor ship *Valeria-Sperrbrecher 176* (1450 tons) sunk. German flak ship *Mewa VII-Vp1269*(112 tons) sunk
30 Aug 44	Afternoon reconnaissance to the Norwegian coast by ten 455 Sqn Beaufighters as part of a force of fifty aircraft from Langham and North Coates. Led by S/Ldr A.L. Wiggins with an escort of Mustang fighters	Some heavy flak encountered from the vicinity of Egero Light and Obrestad Light. No enemy shipping sighted
31 Aug 44	Night anti-shipping patrols along the Dutch coast by six 455 Sqn Beaufighters operating independently	Thunderstorms encountered. No sightings
1 Sep 44	Evening reconnaissance to Den Helder by six 455 Sqn Beaufighters with a Mustang fighter escort	Small fishing vessels seen at the entrance to Den Helder
2 Sep 44	Night patrol along the Dutch Coast by five 455 Sqn Beaufighters	Intense flak from Ostend. No shipping sighted
3 Sep 44	Afternoon combined Langham/North Coates Wing reconnaissance patrol to the Norwegian coast, including twelve 455 Sqn Beaufighters. Led by S/Ldr C.G. Milson with a Mustang fighter escort	A cargo liner was found that Milson judged to be a hospital ship. No attack
6 Sep 44	Morning reconnaissance to Den Helder Harbour by a single 455 Sqn Beaufighter (F/O A.E. Jones RAF and P/O A. Jones RAF)	Many ships reported
6 Sep 44	Evening Langham Wing convoy strike off Heligoland including thirteen 455 Sqn Beaufighters. Led by S/Ldr A.L. Wiggins with a Mustang fighter escort from RAF Coltishall	Swedish merchant vessel *Rosafred* (1496 tons) sunk. Norwegian merchant vessel *Breda* (1261 tons) sunk. German light ship *Emil* (400 tons) sunk
8 Sep 44	Evening combined Langham/North Coates Wing reconnaissance patrol off Stavanger including nine 455 Sqn Beaufighters. Led by W/Cdr Tacon 236 Sqn RAF	Dutch coaster *Hengels* (195 tons) sunk
9 Sep 44	Early morning reconnaissance patrol off the Dutch coast by a single 455 Sqn Beaufighter (W/O J.C. Ayliffe and F/Sgt W.H.H. Iggulden RAF)	Two E-boats reported
9 Sep 44	Morning reconnaissance by the Langham Wing off the Dutch coast including eight 455 Sqn Beaufighters	Large convoy sighted by a 455 Sqn aircraft which was separated from the main force
9 Sep 44	Afternoon reconnaissance patrol off the Dutch coast by a single 455 Sqn Beaufighter (F/O C.G. Thompson and W/O I.M. Gordon)	Eight merchant vessels sighted at Borkum. Aircraft diverted to Little Snoring field
9 Sep 44	Afternoon reconnaissance patrol to Nordernay by a single 455 Sqn	Convoy sighted

	Beaufighter (F/O J.G. Cox and W/O A.W. Ibbotson)	
9 Sep 44	Afternoon reconnaissance patrol along the Dutch coast by a single 455 Sqn aircraft (W/O R. Walker RAF and F/Sgt T. Rabbitts RAF)	Eight vessels sighted at Den Helder
9 Sep 44	Afternoon reconnaissance patrol by a single 455 Sqn Beaufighter (F/O W.M. Stanley and F/O K.A. Dempsey)	Nothing sighted. Aircraft UB-J, NT 892 crashed on return
10 Sep 44	Morning combined Langham/North Coates Wing reconnaissance along the Dutch coast including thirteen 455 Sqn Beaufighters. Led by S/Ldr A.L. Wiggins with Mustang fighter escort	Nothing sighted
10 Sep 44	Evening combined Langham/North Coates wing reconnaissance to the Frisian Islands by 38 Beaufighters, including eight from 455 Sqn. Led by W/Cdr Tacon 236 Sqn RAF with a Mustang fighter escort	Attacks made on shore gun positions
11 Sep 44	Combined Langham/North Coates Wing strike off southern Norway including eight 455 Sqn aircraft. Led by S/Ldr A.L. Wiggins with a Mustang fighter escort	Two minesweepers (M.426 and M.462) sunk and two damaged
12 Sep 44	Combined Langham/North Coates Wing strike at the Marsdiep including ten 455 Sqn Beaufighters. Led by W/Cdr Tacon 236 Sqn RAF with a Mustang fighter escort	Many vessels damaged
13 Sep 44	Evening Langham Wing patrol including ten 455 Sqn Beaufighters. Led by S/Ldr A.L. Wiggins with a Mustang fighter escort	No shipping sighted. Heavy flak from Wangerooge
16 Sep 44	Afternoon patrol to Heligoland by six 455 Sqn Beaufighters, with the North Coates Wing	Twelve unidentified vessels sighted. No attack
17 Sep 44	Early morning reconnaissance by the Langham Wing with Beaufighters from 143 Sqn RAF to the Frisian Islands, with six 455 Sqn Beaufighters. Led by F/Lt J.M. Pilcher	Landing barge and escort vessel sighted
18 Sep 44	Afternoon combined Langham/North Coates Wing reconnaissance including fourteen 455 Sqn Beaufighters. Led by W/Cdr J.N. Davenport with a Mustang fighter escort	Sighted and circled twenty small craft, possibly tank landing craft. No attack made
20 Sep 44	Vehicle accident near Langham	One 455 Sqn ground crew member killed (LAC L.L. Harris)
23 Sep 44	Reconnaissance to Den Helder Harbour by single 455 Sqn Beaufighter (F/O D. Whishaw and F/Sgt W.H.H. Iggulden RAF)	Large volume of shipping reported

23 Sep 44	Combined Langham, North Coates and Strubby Wing strike at Den Helder Harbour including fourteen 455 Sqn Beaufighters. Led by W/Cdr Cartridge 254 Sqn RAF with a Mustang fighter escort from RAF Coltishall	The formation straggled and were eventually led to attack shore installations
25 Sep 44	Combined Langham and North Coates Wing strike at Den Helder including twelve 455 Sqn Beaufighters. Led by S/Ldr Hammond 489 Sqn RNZAF with a Mustang force from Coltishall	German Minesweeper M.471 (750 tons) sunk. Dutch Harbour defence ship *Jannetje- Vs 423* (107 tons) sunk. One 455 Sqn crew lost. (F/O C.E. Cock and W/O A.R. Lyneham in UB-W, NT 987)

S/Ldr C.G. Milson appointed Commanding Officer of 455 Sqn

27 Sep 44	Afternoon reconnaissance by the Langham Wing from Den Helder to Borkum, including six 455 Sqn Beaufighters	No shipping sighted
30 Sep 44	Combined Langham and North Coates Wing reconnaissance, including eleven 455 Sqn Beaufighters. Led by S/Ldr C.G. Milson with a Mustang fighter escort	No shipping sighted
3 Oct 44	Reconnaissance to Schillig Roads and Borkum by a single 455 Sqn Beaufighter (F/O W.G. Herbert and F/Sgt J.W. Tucker)	Two small craft sighted
5 Oct 44	Afternoon reconnaissance off Borkum by single 455 Sqn Beaufighter (P/O J.C. Payne and F/Sgt J. Rennie RAF)	Reported seven coaster vessels
6 Oct 44	Night convoy strike off the Hook of Holland by a force of eight 455 Sqn Beaufighters	Probable hits on three minesweepers
13 Oct 44	Reconnaissance in force by the Langham Wing from Borkum to Nordernay, with ten 455 Sqn Beaufighters. Led by F/Lt Stachberry 489 Sqn RNZAF with a Mustang fighter escort from Coltishall	Three fishing vessels sighted. No attack
15 Oct 44	Morning patrol to the vicinity of the Marsdiep by four 455 Sqn Beaufighters, led by F/O R.C. McColl	Shipping seen in the Marsdiep
15 Oct 44	Morning patrol to Den Helder by a single 455 Sqn aircraft (F/O E.W. Watson and F/O A.V.S. Smith RAF)	Four vessels reported
15 Oct 44	Early morning strike on four armed coastal vessels off the Dutch Coast by six 455 Sqn aircraft. Led by S/Ldr J.M. Pilcher	Dutch Harbour defence vessel *Europa* (339 tons) sunk. German Harbour defence vessel KFK.93 (110 tons) sunk. German light gun barge *Margaret-LAT.15* (200 tons) sunk. One 455 Sqn crew lost after ditching (F/Sgt N.G. Steer and F/O B.T. Roberts in UB-V, NV 431)

15 Oct 44	Afternoon shipping strike in Schillig Roads by the Langham and North Coates wings including fourteen 455 Sqn Beaufighters. Led by W/Cdr A.L. Wiggins with a Mustang fighter escort	254 Sqn attacked a small convoy. A smoke screen was generated from Heligoland. Intense flak from the shore. Three naval auxiliary vessels damaged
19 Oct 44	Morning armed reconnaissance from Texel to The Hague by eleven 455 Sqn Beaufighters, in two formations. First formation led by F/Lt F.L. Proctor, the second by F/Lt E.W. Watson	No shipping found
19 Oct 44	Special reconnaissance patrol by a single 455 Sqn aircraft (F/O J.A. Hakewell and F/Sgt F.G. Sides) in the Heligoland-Borkum area	Convoy sighted, and three Ju-52s attacked. No results seen
22 Oct 44	455 Sqn moves to RAF Dallachy, Scotland	

RAF Dallachy

27 Oct 44	455 Sqn ready for operations from RAF Dallachy with 18 Group Coastal Command	
27 Oct 44	Afternoon patrol from Helliso Light to Stadlandet by ten 455 Sqn Beaufighters and ten Beaufighters from 489 Sqn RNZAF. Led by F/Lt Branton 489 Sqn RNZAF	Patrol abandoned due to poor weather
30 Oct 44	Morning patrol from Stavanger to Utvaer Light by eight 455 Sqn Beaufighters and ten Beaufighters from 489 Sqn RNZAF	No targets sighted
1 Nov 44	Morning reconnaissance from Obrestad Light to Kristiansand by eleven 455 Sqn Beaufighters with eleven Beaufighters from 489 Sqn RNZAF. Led by W/Cdr C.G. Milson	No sightings. Aircraft collision. One 455 Sqn crew lost (F/O G.H. Hammond and F/Sgt G.A. Henry in UB-Y, NT 915). One 489 Sqn crew lost
4 Nov 44	Afternoon reconnaissance patrol from Stavanger to Bergen by eleven 455 Sqn Beaufighters with six Beaufighters from 489 Sqn RNZAF. Led by S/Ldr J.M. Pilcher	Nothing seen
5 Nov 44	Afternoon anti U-boat sweep in the vicinity of Obrestad Light by six 455 Sqn Beaufighters. Led by F/Lt R.W. Kimpton with Outriders from 333 Sqn RAF	Nothing seen
8 Nov 44	Shipping strike at Midtgulen Fiord by the Dallachy Wing with six 455 Sqn Beaufighters. Led by W/Cdr Gadd 144 Sqn RAF	German merchant vessels *Aquila* (3495 tons) and *Helga Ferdinand* (2566 tons) sunk. Norwegian merchant ship *Framnaes* (307 tons) damaged
8 Nov 44	Afternoon anti U-boat patrol in the	Nothing sighted

	vicinity of Stavanger by three 455 Sqn Beaufighters and four Beaufighters from 404 Sqn RCAF. Led by W/Cdr C.G. Milson with a fighter escort	
9 Nov 44	Afternoon anti U-boat patrol by four 455 Sqn Beaufighters with four Beaufighters from 404 Sqn RCAF. Led by S/Ldr Schoales 404 Sqn	Low cloud and poor visibility. Nothing seen
10 Nov 44	Morning reconnaissance patrol by the Dallachy Wing from Utvaer Light to Svinoy Light with six 455 Sqn aircraft	Poor visibility, nothing sighted
15 Nov 44	Armed reconnaissance patrol to the vicinity of Aalesund by the Dallachy Wing, with ten 455 Sqn Beaufighters. Led by W/Cdr Pierce 404 Sqn RCAF with a Mustang fighter escort from 315 (Polish) Sqn	Merchant vessel sighted near Aalesund. No attack made
17 Nov 44	Armed reconnaissance patrol by the Dallachy Wing north from Svinoy Light with eight 455 Sqn Beaufighters. Escorted by Mustang fighters with Outriders from 333 Sqn RAF	Heavy flak encountered. No attack
21 Nov 44	Afternoon anti-shipping patrol to the vicinity of Aalesund by the Dallachy and Banff wings with eleven 455 Sqn aircraft	Small vessels seen
25 Nov 44	Dallachy Wing strike to Haugesund with eleven 455 Sqn Beaufighters. Led by W/Cdr Atkinson 235 Sqn RAF, with 39 Mosquitoes from the Banff Wing	Strike turned back near Haugesund due to poor visibility. Two 455 Sqn aircraft collided. All aircraft returned safely
26 Nov 44	Afternoon anti-shipping patrol by the Dallachy Wing from Kristiansand to Stavanger with five 455 Sqn Beaufighters. Led by S/Ldr Rogers 144 Sqn RAF	Nothing seen
29 Nov 44	Armed reconnaissance from Stadlandet to Molde Fiord by six 455 Sqn Beaufighters with nine Beaufighters from 404 Sqn RCAF	No shipping sighted
5 Dec 44	Afternoon shipping strike at Orsta Fiord by thirteen 455 Sqn Beaufighters with five Beaufighters from 489 Sqn RNZAF. Led by S/Ldr J.M. Pilcher	Two German merchant vessels *Radbod* (4354 tons) and *Albert Janus* (2275 tons) sunk. One 455 Sqn crew lost (F/O J.A. Hakewell and F/Sgt F.G. Sides in UB-V, NV 438) One 455 Sqn aircraft ditched near Sumburgh (UB-M, NT 914). The crew (P/O A.E. Winter and P/O C.J. Dunshea) were rescued
6 Dec 44	Anti-shipping patrol from Froy Fiord to Vaagso by the Dallachy Wing with six 455 Sqn aircraft. Led by W/Cdr Pierce 404 Sqn RCAF with a Mustang fighter escort	Three merchant ships found off Moldoy, but the formation was unable to attack because of the narrowness of the fiord and the intense flak

7 Dec 44	Afternoon patrol to the vicinity of Svinoy Light by the Dallachy Wing with eleven 455 Sqn Beaufighters. Escorted by a Mustang fighter escort and Mosquitoes from the Banff Wing	Six German fighters seen taking off from Gossen airfield. German fighters pursued 489 Sqn RNZAF Beaufighters. Mustang escort engaged fighters. Patrol abandoned
9 Dec 44	Attack on a single merchant vessel off Vilnes Fiord by the Dallachy Wing, with seven 455 Sqn Beaufighters. Led by S/Ldr Duncanson from 144 Sqn RAF with a Mustang fighter escort from 315 (Polish) Sqn	Norwegian merchant vessel *Havda* (678 tons) sunk
11 Dec 44	Patrol by the Dallachy Wing to Norwegian coast departed early afternoon, with five 455 Sqn Beaufighters	Patrol recalled because of bad weather
12 Dec 44	Late morning patrol by the Dallachy Wing from Utvaer Light to Bremanger, including nine 455 Sqn Beaufighters. Led by F/Lt F.L. Proctor with a Mustang fighter escort from 315 (Polish) Sqn	Several small ships seen. No attack
13 Dec 44	Afternoon patrol by the Dallachy Wing from Utvaer Light to Svinoy Light including seven 455 Sqn Beaufighters. Led by F/O C.G. Thompson with a Mustang fighter escort from 315 (Polish) Sqn	Hospital ship sighted, no attack
14 Dec 44	Late morning patrol by the Dallachy Wing from Ytteroerne Light to Utvaer Light including six 455 Sqn Beaufighters. Led by S/Ldr R.W. Kimpton with a Mustang fighter escort	No targets found. Dallachy unfit for landing. Force returned to Banff
18 Dec 44	Afternoon patrol by the Dallachy Wing from Lister Light to Kristiansand including six 455 Sqn Beaufighters	Patrol abandoned due to heavy rain and poor visibility
19 Dec 44	Morning patrol by the Dallachy Wing towards the Norwegian coast including six 455 Sqn Beaufighters. Led by F/Lt Cannell 489 Sqn RNZAF	Cloud base 500 ft. Visibility 1.5 km. Patrol abandoned
23 Dec 44	Morning patrol to the vicinity of Aalesund by ten 455 Sqn Beaufighters and ten Beaufighters from 489 Sqn RNZAF. Led by F/Lt Lynch 489 Sqn	No attack due to poor weather
26 Dec 44	Afternoon anti-shipping patrol along the Norwegian convoy route by the Dallachy Wing from Egero Light to Kristiansand including nine 455 Sqn Beaufighters. Led by F/Lt Cannell 489 Sqn RNZAF	Nothing sighted

28 Dec 44	Afternoon patrol by the Dallachy Wing from Utvaer Light to Stadlandet including five 455 Sqn Beaufighters. Escorted by Mustang fighters from 315 (Polish) Sqn	Frequent heavy showers, no suitable target found
29 Dec 44	Afternoon patrol by the Dallachy Wing from Obrestad Light to Kristiansand including six 455 Sqn Beaufighters. Led by S/Ldr Schoales 404 Sqn RCAF with a Mustang fighter escort from 315 (Polish) Sqn	Nothing sighted
31 Dec 44	Afternoon patrol along the Norwegian convoy route from the vicinity of Stavanger, by the Dallachy Wing including eight 455 Sqn Beaufighters. Led by F/Lt F.L. Proctor with a Mustang fighter escort from 315 (Polish) Sqn	Four neutral merchant ships sighted. No attack
6 Jan 45	Attack on a coastal barge in Fedje Fiord by six 455 Sqn Beaufighters and six Beaufighters from 404 Sqn RCAF. Led by W/Cdr C.G. Milson with a Mustang fighter escort from 315 (Polish) Sqn	Dutch lighter MW.151 (300 tons) sunk
8 Jan 45	Strike force from the Dallachy Wing attacked shipping at Bjorne Fiord. Led by S/Ldr Christison from 404 Sqn RCAF, the force included six 455 Sqn Beaufighters	Norwegian merchant vessel *Fusa* sunk. Norwegian fishing vessel sunk
9 Jan 45	Dallachy Wing strike at Fuglsaet Fiord including eight 455 Sqn Beaufighters. Led by S/Ldr J.M. Pilcher	Norwegian merchant vessel *Sirius* (938 tons) sunk
10 Jan 45	Dallachy Wing strike off Haramso Island including six 455 Sqn aircraft. Led by S/Ldr Christison from 404 Sqn RCAF with a Mustang fighter escort from 315 (Polish) Sqn and outriders from 333 Sqn	One merchant ship sunk, one minesweeper damaged. One 455 Sqn Beaufighter shot down, both crew members killed (P/O A.E. Winter and P/O C.J. Dunshea in UB-M, RD 141)
11 Jan 45	Afternoon patrol by the Dallachy Wing to Flekke Fiord including four 455 Sqn Beaufighters. Led by F/Lt R.C. McColl with a fighter escort	Force attacked by seven German fighters. Patrol abandoned
12 Jan 45	Afternoon patrol by the Dallachy Wing to Utvaer Light including eight 455 Sqn aircraft	Formation searched many fiords. No attacks
14 Jan 45	Dallachy Wing anti-shipping patrol to the vicinity of Stadlandet including six 455 Sqn Beaufighters, with a Mustang fighter escort.	Clouds and poor visibility prevented any attack
15 Jan 45	Dallachy Wing patrol to Sogne Loch including six 455 Sqn Beaufighters.	Snow storms near Bremanger, small vessels seen. No attack made
25 Jan 45	Afternoon patrol by the Dallachy Wing from Sandoy Light to Stadlandet	Low clouds covered mainland. Six small ships seen - not in convoy. No attack

	including seven 455 Sqn Beaufighters	
29 Jan 45	Dallachy Wing patrol from Svinoy Light to Ytteroerne Light including eight 455 Sqn Beaufighters. Led by S/Ldr N.R. Smith with a Mustang fighter escort	Fishing vessels sighted. Heavy snow storm interrupted patrol
3 Feb 45	Late morning patrol by the Dallachy Wing from Utvaer Light to Ytteroerne Light including six 455 Sqn Beaufighters. Led by W/Cdr Gadd 144 Sqn RAF	No suitable targets found
9 Feb 45	Dallachy Wing strike at Forde Fiord including eleven 455 Sqn Beaufighters. Led by W/Cdr C.G. Milson, with a Mustang fighter escort from 65 Sqn RAF	Force attacked by enemy fighters. One crew (F/Lt R.C. McColl and W/O L.L. MacDonald in UB-O, NV 199) taken prisoner. One crew (W/O D.E. Mutimer and F/Sgt J.D. Blackshaw UB-V, NV 196) killed. A *Narvik* class destroyer damaged
10 Feb 45	Dallachy Wing air-sea rescue patrol from Sandoy Light to Ytteroerne Light including five 455 Sqn Beaufighters. Led by W/Cdr Pierce from 404 Sqn RCAF with a Mustang fighter escort from 65 Sqn RAF	No result
16 Feb 45	Afternoon anti-shipping patrol by the Dallachy Wing to Nord Fiord including five 455 Sqn Beaufighters, with a Mustang fighter escort from 65 Sqn RAF	Heavy cloud cover prevented attack
21 Feb 45	Morning anti-shipping patrol by the Dallachy Wing to the Norwegian coast, including eight 455 Sqn Beaufighters. Led by S/Ldr J.M. Pilcher with a Mustang fighter escort from 65 Sqn RAF	Patrol abandoned because of bad weather – visibility 200 metres
23 Feb 45	Night anti-shipping patrol by the Dallachy Wing along the southern Norwegian coast (Kristiansand to Lindesnes) including four 455 Sqn Beaufighters	Nothing sighted
26 Feb 45	Late evening convoy attack in the vicinity of Mandal by the Dallachy Wing including three 455 Sqn Beaufighters	Three vessels damaged including the Norwegian merchant ship *Rogn* (835 tons). One 455 Sqn crew member lost (F/O H.L. Brock in UB-L, NV 414)
2 Mar 45	Three separate night patrols along the Norwegian coast by individual 455 Sqn aircraft	Nothing sighted
4 Mar 45	Afternoon patrol by Dallachy Wing from Utvaer Light to Froy Fiord including six 455 Sqn Beaufighters. Led by W/Cdr Pierce from 404 Sqn RCAF with a Mustang fighter escort from 65 Sqn RAF	Heavy cloud cover prevented attacks
8 Mar 45	Dallachy Wing attack at Midtgulen Fiord including eight 455 Sqn Beaufighters.	Merchant vessels *Moshill* (2960 tons) and *Ex-Larenburg* (3655 tons) damaged.

	Led by S/Ldr J.M. Pilcher with a Mustang fighter escort from 65 Sqn RAF	Danish merchant ferry *Heimdal* (978 tons) damaged. One 455 Sqn crew lost (W/O W.D. Mitchell and F/Sgt I.H. Jury in UB-O, RD 132)
13 Mar 45	Dallachy Wing patrol from Sandoy Light to Svinoy Light including six 455 Sqn Beaufighters. Led by W/Cdr Gadd 144 Sqn RAF	No shipping sighted
17 Mar 45	Afternoon patrol by the Dallachy Wing from Stadlandet to Storholm Fiord, including six 455 Sqn Beaufighters. Led by S/Ldr Schoales from 404 Sqn RCAF with a Mustang fighter escort from 65 Sqn RAF.	Nothing sighted. Clouds prevented the Outriders from searching inland
18 Mar 45	Morning patrol by the Dallachy Wing from Utvaer Light to Svinoy Light including eight 455 Sqn Beaufighters. Led by W/Cdr C.G. Milson	Nothing sighted. Visibility decreasing
23 Mar 45	Night patrol along the southern Norwegian coast including eleven 455 Sqn Beaufighters	Nothing sighted
24 Mar 45	Dallachy Wing strike at Egersund Harbour including eight 455 Sqn Beaufighters. Led by S/Ldr Christison 404 Sqn RCAF with a Mustang fighter escort from 65 Sqn RAF	German merchant vessel *Thetis* (2788 tons) sunk. Norwegian merchant vessel *Sarp* (1116 tons) sunk. One 455 Sqn crew lost (W/O G.E. Longland and F/Sgt E.C. Nayda in UB-X, RD 329)
26 Mar 45	Early morning patrol from Utvaer Light to Ytteroerne Light by six 455 Sqn Beaufighters, led by W/O N.P. Turner	Very poor visibility. Nothing sighted
30 Mar 45	Afternoon patrol from Ytteroerne Light to Svinoy Light by six 455 Sqn Beaufighters and three Beaufighters from 404 Sqn RCAF	Low clouds prevented patrol flying inland. No result
31 Mar 45	Early morning patrol to southern Norwegian coast by six 455 Sqn Beaufighters with two Beaufighters from 144 Sqn RAF	Nothing sighted
2 Apr 45	From Sumburgh, an evening patrol to Storholm Light by six 455 Sqn Beaufighters and four Beaufighters from 489 Sqn RNZAF	Two merchant vessels sighted in Kritiansund (North) Harbour. No attack
3 Apr 45	Dallachy Wing patrol to Egersund Harbour including thirteen 455 Sqn aircraft. Led by W/Cdr C.G. Milson, the strike was cancelled because of poor weather	One 455 Sqn crew member lost when a Beaufighter ditched (W/O W.T. Furlong)
4 Apr 45	Dallachy Wing strike at Aardals Fiord including eight 455 Sqn Beaufighters. Led by F/O Southernwood from 489 Sqn RNZAF with a Mustang fighter	German merchant vessel damaged. One damaged 455 Sqn aircraft escorted to Shetland Islands (UB-F, LZ 407 flown by F/O S.J. Sykes)

	escort from 19 Sqn RAF.	
5 Apr 45	Early morning reconnaissance from Svinoy Light to Aalesund and Nord Fiord by a single 455 Sqn Beaufighter (F/O W.J Edwards and F/Sgt K.L. Hamilton)	Four vessels sighted
5 Apr 45	Dallachy Wing strike to Haried Island including eight 455 Sqn Beaufighters. Led by S/Ldr J.M. Pilcher with Mustang fighter escort from 65 Sqn RAF	Force attacked by German fighters. Two 455 Sqn aircraft collided both crews lost. (F/Lt K.A. Moore and F/Sgt R.E. Bull in UB-C, NT 920, F/O G.W. Hassell and F/O E.J. Loonam in UB-N, NE 207)
7 Apr 45	Dallachy Wing strike at Vadheim Fiord including six 455 Sqn Beaufighters. Led by S/Ldr Duncanson 144 Sqn RAF with a Mustang fighter escort from 19 Sqn RAF and 65 Sqn RAF	German Merchant Vessel *Oldenburg* (4594 tons) sunk. One 455 Sqn aircraft damaged by fighters (UB-G, NE 342 flown by W/O I.A.F. Murray)
11 Apr 45	Dallachy Wing strike in Fede Fiord including six 455 Sqn Beaufighters. Led by S/Ldr Sawyer 489 Sqn RNZAF with Mustang fighter escort from 19 Sqn RAF	German minesweeper M.2 (750 tons) sunk
13 April 45	Three separate reconnaissance patrols along the Norwegian coast by individual 455 Sqn Beaufighters	Small coasters and fishing vessels sighted
14 Apr 45	Dallachy Wing strike in Jossing Fiord including nine 455 Sqn Beaufighters. Led by W/Cdr C.G. Milson with a Mustang fighter escort	Three vessels damaged
19 Apr 45	Late morning reconnaissance from Svinoy Light to Yttereorne Light by a single 455 Sqn Beaufighter (F/O W.J. Edwards and F/Sgt K.L. Hamilton)	Small freighter sighted in Froy Fiord
22 Apr 45	Reconnaissance from Utvaer Light to the vicinity of Sogne Loch by a single 455 Sqn Beaufighter (F/O D.E. Cristofani and W/O W.F.A. Waldock)	Three vessels reported
22 Apr 45	Reconnaissance to the vicinity of Vaagso harbour by a single 455 Sqn aircraft	Merchant vessels reported in Moldoy anchorage
22 Apr 45	Dallachy Wing strike at South Vaagso including five 455 Sqn aircraft. Led by W/Cdr Gadd 144 Sqn RAF with Mustang fighter escort from 19 Sqn RAF and 65 Sqn RAF	German merchant vessel *Elamr* (1000 tons) sunk. One 455 Sqn crew lost (F/O T.J. Higgins and W/O A.J. Mirow in UB-B, NE 444)
23 Apr 45	Dallachy Wing strike at Risnes Fiord including six 455 Sqn aircraft. Led by F/Lt C.G. Thompson with a Mustang fighter escort from 19 Sqn RAF and 65 Sqn RAF	Norwegian merchant vessel *Ingerseks* (4969 tons) sunk
25 Apr 45	Four separate night patrols down the Norwegian coast by single 455 Sqn aircraft	Two unidentified aircraft reported

26 Apr 45	Morning reconnaissance patrol from Svinoy Light to Ytteroerne Light by a single 455 Sqn Beaufighter (F/O J.G. Cox and W/O A.W. Ibbotson)	Several ships sighted
26 Apr 45	Dallachy Wing strike at Fede Fiord including six 455 Sqn aircraft. Led by W/Cdr Gadd 144 Sqn RAF with a Mustang fighter escort from 19 Sqn RAF	Merchant ship *Palmyra* (3600 tons) damaged. One ship sunk. After the strike the Dallachy Wing was attacked by German fighters. F/O W.J. Edwards and W/O K.L. Hamilton shot down and killed in UB-N, RD 429
3 May 45	First-light anti U-boat patrol from Okso Light to Lindesnes by five 455 Sqn Beaufighters. Led by S/Ldr J.M. Pilcher	Nothing sighted

RAF Thornaby

3 May 45	455 Sqn moved to RAF Station Thornaby	
3 May 45	Afternoon strike at Kiel Bay by nine 455 Sqn aircraft. Led by W/Cdr C.G. Milson, with a Mustang fighter escort	Two M-Class minesweepers and a landing craft were damaged
4 May 45	Afternoon patrol by the Dallachy Wing, including thirteen 455 Sqn Beaufighters, departing from North Coates for Kiel Bay. Led by W/Cdr Gadd from 144 Sqn RAF with a Mustang fighter escort	Patrol called off because of poor weather over Denmark

RAF Dallachy

8 May 45	Victory in Europe Day	
11 May 45	Six 455 Sqn Beaufighters escorted the convoy carrying Crown Prince Olaf of Norway. Led by F/O W.G. Herbert	
12 May 45	Six 455 Squadron Beaufighters escorted the convoy carrying Crown Prince Olaf of Norway. Led by W/Cdr C.G. Milson	
21 May 45	Anti U-boat patrol from Lindesnes to Utvaer Light by two 455 Squadron Beaufighters	Nothing sighted
15 May 45	Rocket practice, air-tests and an air-sea rescue patrol	
20 May 45	Notification of Sqn disbandment	W/Cdr C.G. Milson travelled to London to discuss disbandment of the Sqn
22 May 45	All aircraft grounded pending further instructions	
25 May 45	455 Sqn officially disbanded.	

Appendix 2

455 Squadron Organization

This chart shows the organization of 455 Squadron while stationed at RAF Langham and at RAF Dallachy

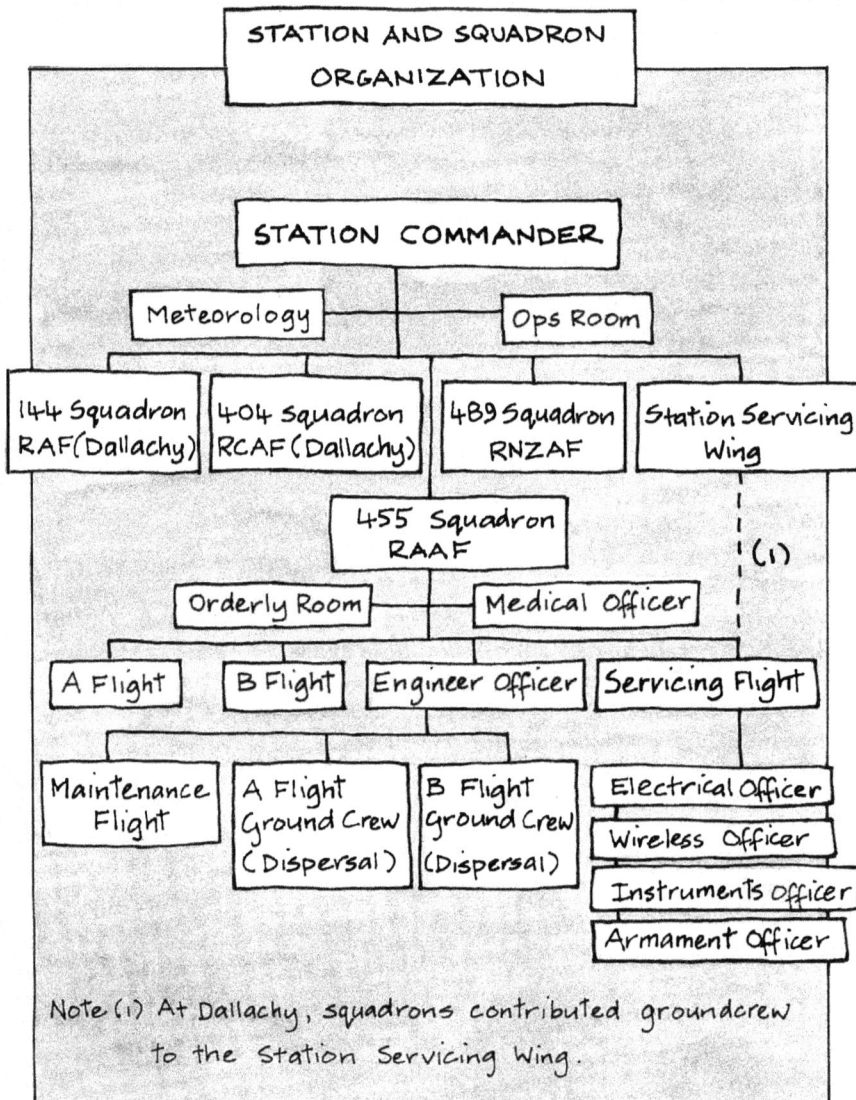

```
                    ┌─────────────────────────────┐
                    │  STATION AND SQUADRON       │
                    │      ORGANIZATION           │
                    └─────────────────────────────┘
                                 │
                    ┌─────────────────────────────┐
                    │      STATION COMMANDER      │
                    └─────────────────────────────┘
            ┌────────────────┐            ┌────────────┐
            │  Meteorology   │────────────│  Ops Room  │
            └────────────────┘            └────────────┘

┌────────────────┐ ┌────────────────┐ ┌──────────────┐ ┌──────────────────┐
│ 144 Squadron   │ │ 404 Squadron   │ │ 489 Squadron │ │ Station Servicing│
│ RAF(Dallachy)  │ │ RCAF(Dallachy) │ │    RNZAF     │ │      Wing        │
└────────────────┘ └────────────────┘ └──────────────┘ └──────────────────┘
                    ┌────────────────┐
                    │ 455 Squadron   │                        (1)
                    │     RAAF       │
                    └────────────────┘
            ┌────────────────┐    ┌────────────────────┐
            │  Orderly Room  │────│  Medical Officer   │
            └────────────────┘    └────────────────────┘

┌───────────┐ ┌───────────┐ ┌───────────────────┐ ┌──────────────────┐
│  A Flight │ │  B Flight │ │ Engineer Officer  │ │ Servicing Flight │
└───────────┘ └───────────┘ └───────────────────┘ └──────────────────┘

┌───────────┐ ┌───────────────┐ ┌───────────────┐ ┌───────────────────┐
│Maintenance│ │   A Flight    │ │   B Flight    │ │ Electrical Officer│
│  Flight   │ │  Ground Crew  │ │  Ground Crew  │ ├───────────────────┤
│           │ │  (Dispersal)  │ │  (Dispersal)  │ │  Wireless Officer │
└───────────┘ └───────────────┘ └───────────────┘ ├───────────────────┤
                                                   │Instruments Officer│
                                                   ├───────────────────┤
                                                   │ Armament Officer  │
                                                   └───────────────────┘
```

Note (1) At Dallachy, squadrons contributed groundcrew to the Station Servicing Wing.

214

Appendix 3

Squadron Manning

This roll of Squadron members includes all those who served in 455 Squadron over the period January 1944 to May 1945. The list was compiled from various sources including RAAF and RAF records and the personal papers of Squadron members. Ranks shown are those held at the end of the War or on leaving the Squadron. Ranks are indicated with the abbreviations used during the War. Errors and omissions are regretted.

W/Cdr   Wing Commander
S/Ldr   Squadron Leader
F/Lt   Flight Lieutenant
F/O   Flying Officer
P/O   Pilot Officer
W/O   Warrant Officer
F/Sgt   Flight Sergeant
Sgt   Sergeant
Cpl   Corporal
LAC   Leading Aircraftman
LACW   Leading Aircraftwoman
AC1   Aircraftman Class 1
ACW1   Aircraftwoman Class 1
AC2   Aircraftman Class 2

AIRCREW

Pilots

W/Cdr J.N. Davenport DSO, DFC and Bar, GM, MID
W/Cdr C.G. Milson DSO and Bar, DFC and Bar
W/Cdr A.L. Wiggins DSO, DFC, MID

S/Ldr R.W. Kimpton DFC
S/Ldr J.M. Pilcher DFC and Bar
S/Ldr N.R. Smith DFC

F/Lt J.S. Addison
F/Lt B. Atkinson RAF (killed on operations 6 May 1944)
F/Lt J.C. Ellis
F/Lt W.G. Herbert
F/Lt J. MacDonald EM
F/Lt I.H. Masson RAF, DFC

F/Lt R.C. McColl DFC
F/Lt K.A. Moore (killed on operations 5 April 1945)
F/Lt C.R. Noble
F/Lt F.L. Proctor DFC
F/Lt A.W. Roper
F/Lt C.G. Thompson DFC and Bar
F/Lt E.W. Watson DFC
F/Lt D. Whishaw DFC

F/O W.M. Barbour (killed on operations 6 July 1944)
F/O J.U. Billing (killed on operations 1 July 1944)
F/O H.L. Brock (killed on operations 26 February 1945)
F/O D.K. Carmody
F/O C.E. Cock DFC (killed on operations 25 September 1944)
F/O E.F. Collaery (killed on operations 29 June 1944)
F/O J.G. Cox DFC
F/O D.E. Cristofani
F/O T.G. Davis
F/O W.J. Edwards (killed on operations 26 April 1945)
F/O L.W. Farr DFC
F/O P.S. Gumbrell RAF (killed in training accident 29 December 1943)
F/O J.A. Hakewell (killed on operations 5 December 1944)
F/O G.H. Hammond (killed on operations 1 November 1944)
F/O H.O. Hanson
F/O G.W. Hassell (killed on operations 5 April 1945)

F/O T.J. Higgins (killed on operations 22 April 1945)
F/O C.S. Howard
F/O W.L. Ives
F/O A.E. Jones RAF
F/O L.A. Kempson (killed on operations 10 August 1944)
F/O G.A.E. Nye
F/O R.H. Penhall
F/O N. Pfeiffer
F/O W.B. Rouse USAAF
F/O B.G. Schinckel
F/O H.R. Spink DFC
F/O W.M. Stanley
F/O S.J. Sykes DSO, DFC
F/O D.F. Wallis
F/O F.O. Williams (killed on operations 8 June 1944)

P/O J.C. Ayliffe DFC
P/O S.J. Cliff RAF
P/O P.L.T. Ilbery
P/O J.R. Maxwell
P/O J.A. McLean
P/O J.C. Payne DFM
P/O C.J. Smith
P/O N.A. Wilson (killed in air accident 1 May 1944)
P/O A.E. Winter (killed on operations 10 January 1945)

W/O G.E. Batchelor RAF (killed on operations 10 August 1944)
W/O R. Dunn
W/O W.T. Furlong (killed on operations 3 April 1945)
W/O W.T. Jones (killed on operations 10 August 1944)
W/O G.E. Longland (killed on operations 24 March 1945)
W/O W.D. Mitchell (killed on operations 8 March 1945)
W/O D.F. Murray
W/O I.A.F. Murray
W/O D.E. Mutimer (killed on operations 9 February 1945)
W/O K.J Newell
W/O N.P. Turner DFC
W/O E.A. Smith
W/O R. Walker DFC, RAF

F/Sgt J. Costello (killed on operations 6 July 1944)
F/Sgt H.T. Doctor
F/Sgt A.C. McNevin

F/Sgt M.M. Roberts (killed on operations 10 July 1944)
F/Sgt N.G. Steer (killed on operations 15 October 1944)
F/Sgt H.A. White
F/Sgt H.H. Williams

Navigators

F/Lt W. Bremner
F/Lt S.E. Drinkwater RAF
F/Lt D.H. Hawkes DFC, RAF
F/Lt M.F.C. Jackson
F/Lt L.M. Jeffreys DFC, MID
F/Lt R.E. Jones DFC, RAF
F/Lt F. Macintyre
F/Lt D.H. Wilks RAF

F/O A.G. Archibald
F/O F.H. Balfour-Ogilvy
F/O W.D. Bawden
F/O J. Belfield RAF
F/O E.F. Bernau
F/O N.E. Blakers
F/O L.R. Clifford DFC
F/O R. Curzon RAF (killed on operations 10 August 1944)
F/O K.A. Dempsey
F/O G.C. Docking
F/O F.G. Dodd RAF (killed on operations 6 July 1944)
F/O J.E. Dovey
F/O T. Dunn
F/O T.O. Edwards (killed on operations 1 July 1944)
F/O F.L. Jack
F/O H.L. Jones RAF
F/O G.T. Kerr RAF
F/O E.J. Loonam (killed on operations 5 April 1945)
F/O C. Mapletoft
F/O T.R. Marrison RAF
F/O C. Mason
F/O W.A. McHugh
F/O R.A.H. Moulton
F/O H.H Osborn RAF
F/O H.W. Pearson DFC
F/O J.F. Peter
F/O A.O. Peters
F/O C.M. Pletoft
F/O B.T. Roberts (killed on operations 15 October 1944)
F/O A.V.S. Smith RAF
F/O M.F. Southgate RAF
F/O D.L. Stanford

F/O L.J. Turner
F/O H.J. Wilding RAF
F/O N.J. Williams

P/O C.J. Dunshea (killed on operations 10
    January 1945)
P/O A. Jones RAF
P/O E.G.A. Miflin
P/O H.R. Morris RAF (killed on operations 10
    August 1944)
P/O J.W. Tucker

W/O D.C. Frey
W/O A.W. Ibbotson
W/O W.H.H. Iggulden RAF
W/O I.M. Gordon DFC
W/O K.L. Hamilton (killed on operations 26
    April 1945)
W/O G.F. Hammond
W/O L.W. Jacobs
W/O R. McGladrigan
W/O A.R. Lyneham (killed on operations 25
    September 1944)
W/O L.L. MacDonald
W/O A.J. Mirow (killed on operations 22 April
1945)
W/O D.A. Mitchell
W/O J.A. Stubbs
W/O W.F.A. Waldock
W/O M.J. Walsh
W/O A.T. Vigor

F/Sgt J.T. Andrew (killed on operations 10 July
    1944)
F/Sgt R.E. Bull (killed on operations 5 April
    1945)
F/Sgt J.D Blackshaw (killed on operations 9
February 1945)
F/Sgt H.E. Brock (killed on operations 10
    August 1944)
F/Sgt B. Crean
F/Sgt R.F. Day RAF
F/Sgt G.A. Henry (killed on operations 1
    November 1944)
F/Sgt J. Heywood RAF
F/Sgt H.J. Hill
F/Sgt M.E. Holder
F/Sgt T.E. Holmes RAF, GM
F/Sgt H.A. Hufford RAF
F/Sgt J. Hyams RAF
F/Sgt H. Jackson
F/Sgt G.A. Johnson
F/Sgt R. Joyner
F/Sgt I.H. Jury (killed on operations 8 March
    1945)

F/Sgt A. Knifton RAF
F/Sgt E.H.E. Knight RAF
F/Sgt J.P Lucy
F/Sgt R.M. Lynes RAF
F/Sgt J. Morris
F/Sgt T.F. Moss RAF
F/Sgt E.C. Nayda (killed on operations 24
    March 1945)
F/Sgt T.G. Phippen RAF
F/Sgt F.B. Power
F/Sgt T. Rabbitts RAF
F/Sgt J. Rennie RAF
F/Sgt W.A. Roach (killed on operations 8 June
    1944)
F/Sgt W. Robertson
F/Sgt F.G. Sides (killed on operations 5
    December 1944)
F/Sgt R. Taylor
F/Sgt J.A. Whitburn (killed on operations 6 Ma
    1944)
F/Sgt W.J White
F/Sgt A.E. Williams
F/Sgt C.A Wilson

Sgt L.H. Bower RAF
Sgt J.A. Haugh RAF
Sgt M.A.C. Maynes RAF
Sgt D.R. Quaintance RAF

Wireless Operator/Air Gunner

W/O F.B. Cardwell
W/O V.C. Locker
W/O K.F MacDonald
W/O C.F. Marshall
W/O J.H. Taylor

F/Sgt G.P Smith
F/Sgt J.S. Smith

GROUNDCREW

F/Sgt A.G. Blundell
F/Sgt H.D.A. Johnson
F/Sgt J.A. McKnight
F/Sgt R.J. McNamara
F/Sgt C. Masters
F/Sgt H.Orr MID
F/Sgt W.H. Waldron

Sgt B.M. Abbott RCAF
Sgt R.E. Beeby RAF
Sgt P.C. Bennett
Sgt R.A. Brayshaw RAF
Sgt A.B. Carter

Sgt N. Cottee
Sgt W.S. Crozier
Sgt E.R. Fenning
Sgt V.J. Goy RAF, MID
Sgt L.W. Green
Sgt J.W. Jones
Sgt J.F. Lefoe
Sgt W. MacDougall
Sgt H.V Perryer
Sgt B.D. Robb RCAF
Sgt H.V. Sands
Sgt J.D. Simpson
Sgt R.A. Tomlin RAF
Sgt A.R. Turner
Sgt H.A. Watters
Sgt K.F. Woodland RAF
Sgt S.F. Young

Cpl R.J. Andrew
Cpl H.A. Arney
Cpl R. Badger
Cpl C. Barrassi
Cpl W. Brewis
Cpl N. Burgoyne
Cpl J. Callaghan
Cpl A.B. Carter
Cpl E.J. Clarke
Cpl T. Clifford
Cpl K.J. Coonan
Cpl J.R. Cowie RAF
Cpl F.J. Crispin RAF
Cpl W. Davison RAF
Cpl I.D.Edgar RAF
Cpl W.E. Edick RCAF
Cpl E.R. Fenning
Cpl J.S. Fields
Cpl J. Fisher
Cpl R.E. Garde
Cpl V.J. Goy RAF, MID
Cpl F.S. Hooker RAF
Cpl M.E Johnson
Cpl H.E. Jones
Cpl R. Jones
Cpl J.C. Kennedy
Cpl R.M. Leask RAF
Cpl R.E.F. McAlister
Cpl S.J. McClure
Cpl W.H. Murphy
Cpl A.P. Piggford
Cpl A. Rees RAF
Cpl Sampson
Cpl H.V. Sands
Cpl N. Scroope
Cpl J.D. Simpson
Cpl C.G. Smith RAF

Cpl W.S. Stewart RAF
Cpl K.W. Tilley RAF
Cpl K. Watson
Cpl T.R. Webb RAF
Cpl W. Weston RAF
Cpl R.J. Wilkinson

LAC H.R. Atkins RAF
LAC R.W. Allen RAF
LAC M.E. Augostin
LAC Beecham
LAC W.A. Boon RAF
LAC R. Boot RAF
LAC T.L. Bowden RAF
LAC C.P. Bragg RAF
LAC J.A. Brannigan RAF
LAC R.V. Bright RAF
LAC D. Brown RAF
LAC W. Brown RAF
LAC C.W. Burgess RAF
LAC J.F. Carlon
LAC W. Castell
LAC W.A. Caroline RCAF
LAC P.L. Causley RAF
LAC L.C. Chapman RAF
LAC R.D. Cleverdon RAF
LAC W. Clifford RAF
LAC J.J. Craddon RAF
LAC J. Cranston RAF
LAC J. Davidson RAF
LAC R.P. Darvill RAF
LAC F. Dean RAF
LAC H.B. Depper RAF
LAC L.W. Dobbie
LAC R. Duncanson RAF
LAC R.H. Ellis RAF
LAC C. Fahey RAF
LAC A.C.J. Gibson RAF
LAC T. Gillespie RAF
LAC F.S. Greenall RAF
LAC T. Gudgeon RAF
LAC L. Harris
LAC T.S. Hay RAF
LAC E.C. Henry
LAC C.J. Hooper RAF
LAC A.W. Hughes RAF
LAC V.T. Hunt
LAC M.W. Johnson
LAC H.E. Jones
LAC L.T. Jones
LAC G.M. Kennedy
LAC R. Knox RAF
LAC R. Leonalo RAF
LAC H.B. Males RAF
LAC M.A. McArthur

LAC E.P. McKinney RCAF
LAC W.C. McLaren RAF
LAC C.F. Martin RAF
LAC W. Mitchell RCAF
LAC R. Morrison RAF
LAC R. Muldoon RAF
LAC W. Opie
LAC E.A. Pearson
LAC K. Ramsay
LAC G.J. Reid RAF
LAC K. Renton RAF
LAC D. Ritchie RAF
LAC J. Rothnie RAF
LACW A.C. Sharp RAF
LAC J.A. Simpson RAF
LAC N. Simpson RAF
LAC D.F. Skinner RAF
LAC E.J. Solomon RAF
LAC G.P. Stiff RAF
LAC J.D. Swanson RAF
LAC G. Teague RAF
LAC H. Toomey
LAC W. Valentine RAF
LAC A.W. Walker RAF
LAC K.G. Walter
LAC A.P. Warren
LAC K.H. Watts RAF
LAC B.E. Wellington
LAC H.A. Willies RAF
LAC J.I. Wooster RAF
LAC L.H. Wylie RAF

AC1 E. Flood RAF
AC1 M.A. McArthur
AC1 J. Miller
AC1 A.E. Price RAF
ACW1 M. Reilly RAF
AC1 J. Swerling RAF

AC2 W. Henderson RAF
AC2 D. Ritchie RAF

SQUADRON STAFF

S/Ldr R.D. Macbeth MID

F/Lt J. Berry
F/Lt W.A. Branch
F/Lt W.D. Baird
F/Lt R. Christie

F/O Hopman

P/O S.D. Champion

W/O C. Masters

F/Sgt J.S. Smith

Sgt L. Bates
Sgt G.C. Bold
Sgt D.C D'Helin
Sgt W.C. Symonds

Cpl R. Gunnel RAF
Cpl R.W. Woolf RAF

LAC A.E. Alder RAF
LAC R. Bartle RAF
LAC R. Benzie RAF
LAC P.J.K. Bluett-Duncan RAF
LAC A. Bryson RAF
LAC C.W. Burscough RAF
LAC S. Charlton RAF
LAC W. Clifford RAF
LAC P.L. Cook RAF
LACW A. Coyle
LAC R. Davidson RAF
LAC J. Diamond RAF
LAC F.L. Downs RAF
LAC R. Evans RAF
LAC F. Eyeley RAF
LAC C.V.S Fairbrother RAF
LAC W. Haggart RAF
LAC A.L. Jackson RAF
LAC S. Kilby RAF
LACW D.V. Male RAF
LAC J. McCabe RAF
LAC J. McFarr RAF
LAC J. Mulready
LAC H. Osborne RAF
LAC T. Patterson RAF
LAC C.R. Porter RAF
LAC R. Sharp RAF
LACW E. Thornhill RAF
LACW M.S. Walker RAF
LACW M. Watson RAF
LACW R. Weir RAF

AC1 H. Deane RAF
ACW1 J. Haggart RAF
ACW1 I.M. Whiting

AC2 W.E. Field RAF

Appendix 4

Honours and Awards

These decorations were awarded to members of 455 Squadron over the period January 1944 to May 1945. Ranks and previous awards are those at the time of the award.

CITATIONS
George Medal

Wing Commander Jack Napier Davenport DSO, DFC

One evening in September 1944, a Beaufighter aircraft, returning from operations with damaged engines, crashed on a runway. The petrol tanks burst and the aircraft became a mass of flames, with bursting cannon shells and ammunition. Wing Commander Davenport, having witnessed the crash, hastened to the scene in his car, leading the ambulance and fire tender. While attempting to subdue the flames with their hoses, the fire crew were forced back by the heat and exploding ammunition. The observer in the aircraft was able to open his hatch and escape by jumping through the fire, but the pilot's cockpit was completely surrounded by flames. His cockpit hatch was seen to open partially and fell shut again. Wing Commander Davenport immediately dashed forward, opened the pilot's top hatch and struggled to free him. The pilot was severely burnt and his feet were jammed. Wing Commander Davenport pulled him out of his flying boots and lifted him bodily through the blazing inferno to safety. Wing Commander Davenport sustained shock, with burns to his face, hair and hands. By his prompt and most courageous action, he saved the pilot's life.

Flight Sergeant Thomas Edward Holmes RAF

One night in May, 1944, Flight Sergeant Holmes was observer in a Beaufighter aircraft detailed for bombing practice. Soon after the take off one engine failed and, when a forced landing was attempted at a nearby airfield, the aircraft crashed in a wood and burst into flames. Flight Sergeant Holmes was burnt about the face and his harness was on fire before he was able to fight his way out of his compartment. Trees had fallen over the aircraft and were blazing furiously. The airman climbed first along the port side and then along the starboard side of the aircraft in an attempt to rescue his pilot, but was prevented from doing so by the burning trees and the fire in the aircraft. Flight Sergeant Holmes then forced his way through the woods to the front of the Beaufighter but again was unable to get to his companion because of the dense and blazing undergrowth. This airman displayed great courage and persistence in his gallant attempt to rescue the pilot in the face of exploding cannon shells, practice bombs and petrol tanks which had been torn from the aircraft and were blazing nearby. In August, 1944, Flight Sergeant Holmes was again involved in a crash of another aircraft in which he was flying as observer. Despite severe cuts to his head and face he quickly left his compartment, climbed up the wing of the blazing aircraft, and helped his pilot from the cockpit, thereby enabling him to escape with only minor burns.

Bar to Companion of The Distinguished Service Order

Wing Commander Colin George Milson DSO, DFC

Since being awarded the Distinguished Service Order, Wing Commander Milson has participated in many sorties including numerous attacks on enemy shipping. In February, 1945, he led a large formation of aircraft in attacks against an enemy naval force in a fiord in Norway. The vessels were at anchor close in to the shore. Although mountains rise in close proximity to these narrow waters, Wing Commander Milson employed good tactics leading his formation into the attack with great determination. Enemy fighter opposition and considerable anti-aircraft opposition were encountered, but the attack was well pressed home. In this brilliantly executed operation, Wing Commander Milson showed skill, courage and devotion to duty of a high order.

Companion of the Distinguished Service Order

Wing Commander Jack Napier Davenport DFC

This officer has taken part in a large number of sorties, including many attacks on shipping, during which much loss has been inflicted on the enemy. Recently he led a formation of aircraft in an attack on an enemy convoy which was escorted by 16 armed ships. In spite of fierce opposing fire a most determined and successful attack was made. Two medium sized merchant vessels and one of the escorting vessels were very severely damaged. In this well executed operation, Wing Commander Davenport displayed high powers of leadership, great skill and determination which contributed materially to the results obtained. This officer has rendered great service and his sterling qualities have impressed all.

Wing Commander Colin George Milson DFC

Since the award of a Bar to the Distinguished Flying Cross, Wing Commander Milson has completed numerous operational missions which have included a number of outstanding attacks in the face of the heaviest opposition. In July and August, 1944, he led a formation of aircraft to attack several strongly escorted enemy convoys and despite adverse weather and heavy defences, the attacks caused much damage and destruction. On one occasion, his aircraft sustained severe damage, but with cool courage and determination, he attacked the vessels and then flew safely to base, a distance of three hundred miles on one engine. In September, 1944, this officer was deputy leader of a force detailed to attack shipping in a heavily defended harbour. While under fire, he carefully manoeuvred his section and in the face of a fierce opposition, led them into attack against heavily armed escort vessels. Two of the attacks were pressed home to deck level. Wing Commander Milson's brilliant leadership, gallantry and careful planning have contributed greatly to the success of many missions.

Flying Officer Stephen Joseph Sykes DFC

This officer has completed a large number of sorties including many attacks on enemy shipping during which he has shown the highest standard of skill and resolution. His example of courage was amply demonstrated on one occasion in April, 1945, when he led the Squadron in an attack on a target at anchor in the South Fiord. Whilst pressing home his attack, Flying Officer Sykes was severely wounded, being affected in the arm and leg. Despite this, he completed his attacking run. Afterwards, Flying Officer Sykes was given first aid. Although suffering acutely he flew his aircraft to base and executed a perfect landing. This officer set a magnificent example of courage, fortitude and devotion to duty.

Bar to Distinguished Flying Cross

Wing Commander Jack Napier Davenport DSO, DFC

Since being awarded the Distinguished Service Order, Wing Commander Davenport has taken part in numerous attacks on enemy shipping. Much of the success obtained can be attributed to this officer's brilliant and courageous leadership. Wing Commander Davenport's outstanding ability and personal example have impressed all.

Squadron Leader Colin George Milson DFC

Since being awarded the Distinguished Flying Cross, this officer has completed numerous attacks on enemy shipping. He has led his striking forces with the greatest dash and his example has been most inspiring. In June, 1944, he flew the leading aircraft of a section which attacked a number of minesweepers near the Hook of Holland. In the face of intense anti-aircraft fire the attack was pressed home with exceptional resolution. His contempt for danger has been a noteworthy feature throughout his tour.

Squadron Leader John Milton Pilcher DFC

Within recent months, Squadron Leader Pilcher has participated in a number of attacks on enemy shipping and has displayed the greatest skill and daring in pressing home his attacks. On two occasions in December 1944, he led the Squadron in attacks on shipping in enemy waters with good results. He is an outstanding pilot whose great courage and devotion to duty have impressed all.

Flight Lieutenant Ralph Edward Jones DFC, RAFVR

This officer is a navigator of outstanding ability who has materially contributed to the successful completion of many sorties. In February, 1945, he was leading navigator of a force which attacked an enemy destroyer and 10 escort vessels in Forde Fiord. Despite the intense anti-aircraft fire, Flying Officer Jones was able to deliver a most accurate account to his pilot whilst the attack was in progress. Again, in September, 1944 Flying Officer Jones gave his pilot valuable assistance during an attack against enemy shipping in Den Helder harbour and obtained some excellent photographs. Throughout his operational career this officer has displayed a fine fighting spirit and great devotion to duty.

Flight Lieutenant Clive Gordon Thompson DFC

This officer has participated in a large number of sorties, including 14 attacks on enemy shipping. In these operations he has displayed great skill and has invariably pressed home his attacks with resolution. As a result he had been responsible for inflicting much damage on the enemy. In April, 1945, Flight Lieutenant Thompson led a formation of aircraft in an attack against an enemy merchantman, escorted by three other vessels, in the narrow Risnes Fiord. Although the approach to the target was over difficult terrain, a well executed attack was made. This officer has displayed commendable courage and devotion to duty.

Distinguished Flying Cross

Squadron Leader Neil Reginald Smith

This officer had completed very many attacks on enemy shipping. On one occasion, in June, 1944 he pressed home an attack against two vessels escorting an important convoy off the Frisian Islands. The opposition by the enemy was very severe, but the photographic evidence brought back by Squadron Leader Smith showed how successful his attacks were. On numerous sorties his aircraft has been badly damaged by enemy fire but he always pressed home his attack. His skilful leadership together with his efficiency and courage, have won great praise.

Squadron Leader Albert Lloyd Wiggins DSO

This officer had participated in numerous sorties, including several successful attacks on enemy shipping. In July, 1944, Squadron Leader Wiggins flew the leading aircraft of a formation which attacked a convoy of ten enemy ships. In spite of considerable anti-aircraft fire the attack was well pressed home. A merchantman was hit and several other vessels set on fire. By his gallant and skilful leadership, Squadron Leader Wiggins played a prominent part in the success of a well executed operation.

Flight Lieutenant Samuel Edwin Drinkwater RAFVR

Flight Lieutenant Drinkwater has been engaged on operations with this Squadron for a considerable time. As navigator, he has participated in a large number of anti-shipping sorties, many of them against important convoys. In May, 1944, as deputy leader of a wing operation, he attacked a convoy of seven enemy merchant ships and 12 escort vessels, causing great damage. In July, 1944, he took part in a similar operation, despite severe opposition and gave his pilot valuable assistance on the homeward flight. He has always displayed outstanding courage and determination.

Flight Lieutenant Michael Frederick Carew Jackson

As navigator, this officer has taken part in a large number of operational sorties against enemy shipping. He has always displayed outstanding navigational ability, courage and great determination. Neither enemy opposition nor adverse weather has ever deterred Flight Lieutenant Jackson from completing his allotted tasks and he has secured some excellent photographs. In September, 1944, this officer participated in an attack against the heavily defended Den Helder harbour. Despite the fierce anti-aircraft fire he obtained an outstanding photograph, in addition to engaging several gun positions and giving his pilot an excellent running commentary. More recently in April, 1945, Flight Lieutenant Jackson was leading navigator of a force detailed for an attack against enemy shipping, the force was intercepted by enemy fighters and his aircraft was singled our for attacks. Largely owing to his excellent directions to his pilot for evasive action and fine handling of his rear gun, the enemy fighters were beaten off.

Flight Lieutenant Robert Webb Kimpton

This officer is a capable and courageous pilot. He has taken part in very many anti-shipping sorties during which several enemy vessels of varying types have been severely damaged. The coolness, initiative and balanced judgment of Flight Lieutenant Kimpton have been most commendable.

Flight Lieutenant Robert Charles McColl

Flight Lieutenant McColl has led the Squadron and, frequently the wing on many daring sorties against enemy shipping. In February, 1945, he was deputy wing leader in a force which attacked an enemy convoy in Forde Fiord. Despite intense anti-aircraft fire from ships and shore batteries he completed a most determined attack on two anti-aircraft ships. On many other occasions, this officer has caused damage and destruction to enemy vessels. He has frequently encountered intense and accurate opposition but this has never deterred him from completing his allotted tasks with courage and resolution.

Flight Lieutenant John Milton Pilcher

This officer has completed many sorties on his second tour of operations during which he has participated in several attacks on shipping. Recently he took part in an attack on a convoy of 19 ships, pressing home his attack with exceptional skill and vigour. Some days later, Flight Lieutenant Pilcher participated in an attack on another convoy. In the face of intense opposing fire, he led his section with great skill and gallantry and his brilliant work contributed in good measure to the success achieved. This officer has set an example of the highest order.

Flight Lieutenant Frank Livingstone Proctor

This officer is a most efficient pilot and captain. He has taken part in a large number of sorties and throughout has displayed a high degree of determination and devotion to duty. One of his achievements was a most damaging attack on an enemy submarine one night in March, 1945. On numerous other occasions, Flight Lieutenant Proctor has executed telling attacks on enemy shipping.

Flight Lieutenant Edward Walter Watson

Flight Lieutenant Watson has completed very many sorties including a number of successful attacks on enemy shipping. On one occasion he observed a force of 15 enemy vessels off the Hook of Holland and, after a message reporting the position had been transmitted he went in to attack. In spite of much fire from the convoy, he made three runs over the target and effectively attacked two of the vessels. In November 1944, he participated in a successful attack on three enemy merchant vessels lying in a fiord surrounded by high hill. In this operation his leadership was brilliant and played a good part in the success achieved. Throughout his tour this officer has displayed the highest standard of courage, skill and determination

Flying Officer Harold Spink and Flying Officer Lloyd Clifford

Flying Officer Clifford and Flying Officer Spink were navigator - wireless operator and pilot respectively of one of a formation of aircraft detailed to attack on enemy naval force consisting of a medium sized naval vessel and nine other armed ships at an anchorage in a narrow fiord.
The formation was met with fierce anti-aircraft fire and as Flying Officer Spink pressed home his attack, a shell burst through the starboard window wounding him in the chest, ribs and arm. Although bleeding profusely he coolly climbed to cloud cover when warned of the presence of enemy fighters. Flying Officer Clifford had also been wounded once by shrapnel entering his leg but nevertheless, after taking a number of photographs of the attack, he went to the assistance of his pilot to whose injured arm he applied a tourniquet. Although growing steadily weaker, Flying Officer Spink flew the aircraft to base. During the homeward flight, Flying Officer Clifford proved his skill by accurate navigation in the face of great difficulties. When coming in to land, it was found that the hydraulic gear had been damaged and the undercarriage could not be lowered. Flying Officer Spink was unable to attempt another circuit of the airfield as his arm was powerless but in spite of this, he effected a successful crash landing even though the port engine caught fire when the aircraft was down to a few hundred feet. This officer and his able, devoted navigator displayed the finest qualities of skill, courage and fortitude on this, their first operational sortie.

Flying Officer Jack Geoffrey Cox

This officer has completed a large number of operational missions, including many valuable reconnaissances during which he has obtained information of great value; he has also taken part in numerous successful attacks on enemy shipping. On one occasion, early in his operational career, Flying

Officer Cox piloted an aircraft in an attack against four enemy minesweepers. During the run-in, his aircraft was hit. Flying Officer Cox was struck in the arm by a piece of shrapnel. Despite this he executed his attack and afterwards flew his badly damaged aircraft back to an airfield where he effected a successful crash-landing. This officer has set a fine example of gallantry and devotion to duty.

Flying Officer Colin Edwin Cock

Flying Officer Cock has participated in a large number of sorties including many attacks on enemy convoys. He is a skilful and courageous pilot who has successfully accomplished his operational sorties regardless of heavy enemy opposition and adverse weather.

Flying Officer Lloyd Wilbur Farr

This officer has participated in very many sorties, including several successful attacks on enemy shipping. In the last of these a convoy of ten merchantmen, escorted by a number of armed ships was attacked. Intense anti-aircraft fire was encountered but although his aircraft was hit, Flying Officer Farr pressed home his attack with the greatest resolution and completely silenced the guns on the vessels at which he directed his fire. His determination and devotion were characteristic of that he has shown throughout his tour.

Flying Officer Ralph Edward Jones RAFVR

This officer has participated in numerous sorties, including several successful attacks on enemy shipping. He is a gallant and devoted member of aircraft crew, whose navigational ability has contributed in good measure to the successes obtained. His keenness, thoroughness and appreciation of the responsibilities entrusted to him have set a very fine example.

Flying Officer Ian Hamilton Masson RAFVR

Now on his second tour of operations this officer has taken part in several attacks on enemy shipping and has displayed courage and skill of a high order. In May, 1944, Flying Officer Masson participated in an attack on a large and heavily armed convoy. Despite intense anti-aircraft fire, this officer pressed home his attack with the greatest resolution. His aircraft was repeatedly hit and sustained much damaged but he flew it back to base where he effected a successful crash landing. His determination on this occasion was typical of that he has shown throughout his tour.

Flying Officer Horace William Pearson

As navigator, this officer has taken part in a good number of sorties, including several attacks on enemy shipping. On the first of these, Flying Officer Pearson's aircraft was badly damaged by enemy fire and had to be brought down onto the sea. After being adrift in a dinghy for eleven hours, Flying Officer Pearson was rescued. He soon resumed operational flying . In April, 1945, he took part in an attack against several enemy minesweepers. During the operation, the aircraft was hit and his pilot was badly injured. With great promptitude, Flying Officer Pearson rendered most efficient first aid and afterwards greatly assisted his injured pilot in his efforts to fly the aircraft home. On reaching base, Flying Officer Pearson manipulated the throttles and hydraulic gear and thus helped to bring the aircraft down safely. This officer has invariably displayed a high degree of skill, determination and devotion to duty.

Flying Officer Matthew Francis Southgate RAFVR

This officer has taken part in many sorties, and has displayed skill and devotion to duty throughout. In July, 1944, he was the navigator of one of a large formation of aircraft which attacked a big enemy convoy. The attack was pressed home with great determination and Flying Officer Southgate rendered valuable service in the co-ordination of a most successful attack. He is a highly skilled and courageous member of aircraft crew.

Flying Officer Stephen Joseph Sykes

Since joining the Squadron, Flying Officer Sykes has taken part in many operational missions, including numerous attacks on enemy shipping. In September, 1944, he took part in an attack on shipping in the harbour at Den Helder. In spite of fierce fire from the enemy's defences this officer pressed home his attack from mast height, obtaining hits on a trawler. His aircraft was repeatedly hit and much damage sustained. A large hole was torn in the nose of the aircraft on the starboard side. Nevertheless Flying Officer Sykes flew back to base. After landing it was discovered that some 3 feet of the top of a ship's mast was embedded in the torn nose of the aircraft. Flying Officer Sykes is an outstanding pilot whose brilliant marksmanship and great determination have earned him much success.

Flying Officer Clive Gordon Thompson

This officer has displayed great skill both as navigator and pilot. In the latter capacity, Flying Officer Thompson had participated in a number of attacks on enemy shipping. In September, 1944, he piloted an aircraft in an attack on shipping at Den Helder. In the face of intense anti-aircraft fire, Flying Officer Thompson attacked three vessels. He displayed a high degree of skill, courage and resolution throughout.

Flying Officer David Whishaw

In September, 1944, this pilot took part in an attack on shipping, in the heavily defended harbour of Den Helder. In spite of concentrated heavy and light anti-aircraft fire he pressed home a devastating attack from very close range. He has participated in many operations and has proved himself a gallant and skilful pilot who can always be relied upon to put up an excellent show.

Pilot Officer Charles Joseph Lewis Smith

This officer had taken part in very many sorties, including numerous attacks on enemy shipping. In August, 1944, he took part in an attack on a well defended enemy convoy. In the fight, Pilot Officer Smith obtained a hit on a medium sized vessel, which caught fire. In September, 1944, he took part in an attack on shipping at Den Helder. In the face of considerable anti-aircraft fire, Pilot Officer Smith attacked shore batteries and afterwards pressed home an attack on a mine-sweeper. Although his aircraft sustained damage making it difficult to control, he flew it safely to base. This officer has set a fine example of courage and tenacity.

Warrant Officer John Colin Ayliffe

Warrant Officer Ayliffe has rendered excellent service. He has completed very many sorties, including numerous attacks on enemy shipping. In these operations he has displayed the greatest keenness to engage the enemy and has invariably pressed home his attacks with the utmost determination. On

226

several occasions his aircraft has been badly damaged by enemy fire but each time he has flown back to base. Warrant Officer Ayliffe has set a splendid example of devotion to duty.

Warrant Officer Ivor Montague Gordon

Warrant Officer Gordon has taken part in a good number of sorties, including numerous attacks against enemy shipping. In April, 1945, he was the navigator in an aircraft detailed to attack shipping in the narrow Risnes Fiord. During the operation the aircraft was hit. The cannon installations caught fire. Although Warrant Officer Gordon promptly used the extinguishers the flames spread. Undeterred, this crew member worked strenuously to get the fire under control. Finally, he succeeded. In so doing he had not hesitated to beat out the flames with only his gloved hands. He afterwards navigated the aircraft to base. His courage and determination were typical of that which he had shown at all times.

Warrant Officer Noel Paige Turner

Warrant Officer Turner has displayed great skill and daring in attacks against enemy shipping. In April, 1945, he took part in an attack against shipping in the anchorage of Moldoy in Sor Vaagso. In spite of fierce fire from the vessels and from land based batteries, Warrant Officer Turner pressed home a most determined attack. He displayed the highest standard of bravery and devotion to duty throughout.

Warrant Officer Ralph Walker RAFVR

Warrant Officer Walker has completed many operational and reconnaissance flights since joining 455 Squadron. In May, 1944, he participated in an attack on a convoy of 19 enemy ships off the Dutch coast. In the face of severe and accurate anti-aircraft fire he pressed home his attack to deck level, although his aircraft was badly damaged. On several other occasions in spite of intense anti-aircraft fire he has pressed home his attack and brought back photographs indicating great destruction.

Distinguished Flying Medal

Flight Sergeant John Charles Payne

Flight Sergeant Payne has participated in numerous sorties including several attacks on shipping. He is a highly efficient and devoted member of the crew. On a recent occasion when nearing base at the conclusion of a sortie, his aircraft commenced to vibrate violently. Soon afterwards it went into a spiral dive but Flight Sergeant Payne succeeded in regaining control. It was then discovered that the starboard elevator had broken away and was suspended in airflow by a control wire. It seemed as though the aircraft would have to be abandoned as once more control was lost and the aircraft went into a steep dive. Flight Sergeant Payne succeeded in levelling out and as there seemed slight improvement in control he resolved to try and save the aircraft. Seven times in all the aircraft dived but this intrepid pilot succeeded in reaching base where he effected a masterly landing. Flight Sergeant Payne set an example of the highest order.

Mentioned in Dispatches

Wing Commander J.N. Davenport	Wing Commander A.L. Wiggins
Squadron Leader L.M. Jeffrey	Squadron Leader R.D. Macbeth
Flight Sergeant H. Orr	Corporal V.J. Goy RAF

Appendix 5

455 Squadron Beaufighters

BEAUFIGHTERS FLOWN BY 455 SQUADRON
DECEMBER 1943 TO MAY 1945*

Aircraft Letter	Aircraft Number	Remarks
UB-A1	N 260	Delivered with the first batch of Beaufighters to arrive on 9 Dec 43.
UB-A	NE 200	Crashed into the sea 8 Jun 44 after engine trouble, F/O F.O. Williams and F/Sgt W.A. Roach killed.
UB-A	NE 760	Crashed into the sea during a patrol to Dutch coast on 10 Jul 44. F/Sgt M.M. Roberts and F/Sgt J.T. Andrew killed.
UB-A	NT 954	Delivered 7 Aug 44. Flown by F/O J.G. Cox on 455 Sqn's last shipping strike for the War on 3 May 45.
UB-B	NE 202	Delivered with the first batch of Beaufighters to arrive on 9 Dec 43.
UB-B	NE 444	Delivered 25 Sep 44. Shot down during strike at South Vaagso on 22 April 1945. F/O T.J. Higgins and W/O A.J. Mirow killed.
UB-C	NE 326	Delivered 18 Dec 43.
UB-C	NE 204	Delivered with the first batch of Beaufighters to arrive on 9 Dec 44. Damaged after belly-landing 25 Mar 44.
UB-C	NE 773	Crashed and burned during formation practice from Langham on 1 Jul 44. The crew F/O J.U. Billing and F/O T.O. Edwards were killed.
UB-C	NT 920	Delivered 29 Aug 44. Lost after colliding with Beaufighter UB-N off the Norwegian coast on 5 Apr 45. F/Lt K.A. Moore and F/Sgt R.E. Bull killed.
UB-C	RD 332	Flown by P/O P.L.T. Ilbery on 455 Sqn's last operational sortie, an anti U-boat patrol on 21 May 45.
UB-D	KW 277	Delivered 25 Jul 44. Shot down during a strike at Heligoland on 10 August 1944. F/O L.A. Kempson and F/O R. Curzon RAF killed.
UB-D	NE 325	Delivered 18 Dec 43.
UB-D	NV 432	Delivered 28 Nov 44.
UB-E	NE 543	Delivered 14 Aug 44. Last flown on 9 Oct 44.
UB-E	NT 947	Delivered 5 Sep 44.
UB-F	LZ 407	Delivered 22 Dec 43. Crash-landed at RAF Leuchars on 6 Jan 44 while practising overshoot procedure. Damaged during a strike at Aardals Fiord on 4 April 1945.
UB-F	NE 348	Delivered 23 Dec 43. Crashed into the sea after being hit by flak during strike off the Dutch coast on 6 Jul 44. F/O W.M. Barbour and F/O F.G. Dodd RAF killed.
UB-F	NT 963	Crash-landed at Langham by F/O S.J. Sykes after a combined wing strike at Den Helder on 12 Sep 44.
UB-G	NE 342	Delivered 23 Dec 43.
UB-G	NE 682	Delivered 14 Sep 44.
UB-G	RD 427	Flown by W/O I.A. Murray on 455 Sqn's last operational sortie, an anti U-boat patrol on 21 May 45.

UB-H	LZ 192	Delivered 22 Dec 43. Ditched after a convoy strike on 29 June 1944. The pilot, F/O E.F. Collaery was killed.
UB-H	NV 423	Delivered 10 Oct 44.
UB-H	NT 959	Shot down during a strike at Heligoland on 10 August 1944. W/O G.E. Batchelor RAF and P/O H.R. Morris RAF killed.
UB-J	LZ 402	Delivered 22 Dec 43. Crash-landed at Langham by F/O J.G. Cox after convoy strike on 29 June 1944.
UB-J	NT 892	Delivered 26 Aug 44. Crash-landed and caught fire at Langham on 9 Sep 44 after a reconnaissance flight.
UB-J	NE 682	
UB-J	NE 347	Delivered 17 Dec 44.
UB-K1	NV 177	
UB-L	NE 340	Delivered 17 Dec 43. Shot down during a strike at Heligoland on 10 August 1944. W/O W.T. Jones and F/Sgt H.E. Brock killed.
UB-L	NV 414	Delivered 29 Nov 44. Aircraft abandoned with engine trouble after a convoy strike near Mandal on 26 Feb 45. Pilot F/O H.L. Brock parachuted but was blown out to sea and drowned.
UB-L	RD 396	Delivered 13 Feb 45. Flown by P/O J.A. McLean on 455 Sqn's last shipping strike for the War on 3 May 45.
UB-M	NE 812	Delivered 10 May 44.
UB-M RAF	NE 206	Delivered with the first batch of Beaufighters to arrive on 9 Dec 43. Failed to return from a strike off the Dutch Coast on 6 May 1944. F/Lt B. Atkinson and F/Sgt J.A. Whitburn killed.
UB-M	NT 914	Delivered 15 Jun 44. Ditched off Shetland Islands (Sumburgh) after being hit by flak during an attack at Orsta Fiord on 5 Dec 44. P/O A.E. Winter and P/O C.J. Dunshea rescued.
UB-M	RD 141	Delivered 28 Dec 44. Aircraft crashed into the sea after attack off Haramso Island on 10 Jan 45. P/O A.E. Winter and P/O C.J. Dunshea killed.
UB-N	NE 207	Delivered with the first batch of Beaufighters to arrive on 9 Dec 43. Collided with Beaufighter UB-C after being attacked by enemy fighters off the Norwegian coast on 5 Apr 45. F/O G.W. Hassell and F/O E.J. Loonam killed.
UB-N	RD 429	Crashed and exploded near Fede Fiord after a strike on 26 April 1945. F/O W.J. Edwards and W/O K.L. Hamilton killed.
UB-O	NE 326	Delivered 18 Dec 43. Crashed and caught fire on landing at Langham on 24 Aug 44 after a morning reconnaissance. Crew (W/O J.A. McLean and F/Sgt W.A. McHugh) unhurt.
UB-O	NV 199	Delivered 5 Sep 44. Shot down during a strike at Forde Fiord on 9 February 1945. F/Lt R.C. McColl and W/O L.L. MacDonald captured.
UB-O	RD 132	Delivered 24 Feb 45. Lost during a strike at Midtgulen Fiord on 8 March 1945. W/O W.D. Mitchell and F/Sgt I.H. Jury killed.
UB-O	RD 461	Flown by F/Lt J.C. Ellis on 455 Sqn's last operational strike for the War on 3 May 45.
UB-P	NE 196	Delivered 18 Dec 43.
UB-P	NV 524	Flown by F/O W.L. Ives on 455 Sqn's last shipping strike for the War on 3 May 45.
UB-Q	NE 798	Delivered 27 Mar 44. First flew on operations on 21 April 1944. Remained in 455 Sqn until April 1945.
UB-R	LZ 537	Delivered 15 Sep 44. Crashed after take-off on 23 Sep 44.
UB-R	NV 424	Flown by S/Ldr N.R. Smith on 455 Sqn's last shipping strike for the War on 3 May 45.
UB-S	LZ 409	Delivered 22 Dec 43. Normally flown by W/Cdr C.G. Milson. This aircraft was retired on 2 Apr 45 after 400 hours flying with 455 Sqn.
UB-S1	NE 669	
UB-T	LZ 407	Delivered 22 Dec 43. Crash-landed at Langham on 28 Jul 44.

UB-T	NE 777	Delivered 30 Oct 44. Belly-landed at Dallachy on 11 Apr 44 following a strike at Fede Fiord.
UB-U	LZ 194	Delivered 15 Dec 43. Hit by flak and ditched off the Dutch coast on 6 Jul 44. One crew member, F/Sgt R. Taylor taken prisoner. The pilot, F/Sgt J. Costello was killed.
UB-U	NE 750	Delivered 29 Sep 44. Flown by W/Cdr C.G. Milson on 455 Sqn's last shipping strike for the War on 3 May 45.
UB-V	NV 431	Ditched after being hit by flak during a strike off the Dutch Coast on 15 October. F/Sgt N.G. Steer and F/O B.T. Roberts killed.
UB-V	NV 196	Shot down during strike at Forde Fiord on 9 February 1945. W/O D.E. Mutimer and F/Sgt J.D. Blackshaw killed.
UB-V	NE 774	
UB-V	NV 438	Delivered 28 Nov 44. Hit by flak and crashed during a strike at Orsta Fiord on 5 December 1944. F/O J.A. Hakewell and F/Sgt F.G. Sides killed.
UB-W	NE 353	Delivered 13 Jan 44. Crashed after take-off on 1 May 44. Pilot P/O N.A. Wilson killed.
UB-W	NT 987	Shot down during a combined wing strike at Den Helder Harbour on 25 September 44. F/O C.E. Cock and W/O A.R. Lyneham killed.
UB-W	NV 543	Delivered 8 Dec 44.
UB-W	NE 353	
UB-X	NE 775	Delivered Mar 44.
UB-X	NE 776	Belly-landed Langham on 15 Jun 44.
UB-X	RD 329	Shot down during a convoy strike at Egersund Harbour on 24 Mar 45. W/O G.E. Longland and F/Sgt E.C. Nayda killed.
UB-X	NE 431	
UB-X	RD 331	
UB-X	NV 450	
UB-Y	NT 915	Delivered 24 Jun 44. Collided with a Beaufighter from 489 Sqn RNZAF and crashed into the sea off the Norwegian Coast on 1 November 1944. F/O G.H. Hammond and F/Sgt G.A. Henry killed.
UB-Y	NE 223	Arrived 5 Feb 44.
UB-Y	RD 459	Flown by F/O R.H. Penhall on 455 Sqn's last shipping strike for the War on 3 May 45.
UB-Z	NE 668	Delivered 16 Mar 44. Ditched off the Dutch Coast on 13 June 1944. F/O D.K. Carmody and F/O G.C. Docking taken prisoner.
UB-Z	NT 923	Delivered 15 Jan 45. Flown by F/O D.E. Cristofani on 455 Sqn's last shipping strike for the War on 3 May 45.
	LZ 193	Delivered 22 Dec 43. Crashed on a demonstration flight on 29 Dec 43. F/O Toombs (489 Sqn RNZAF) and F/O P.S. Gumbrell RAF killed.

★ This list has been compiled from the 455 Squadron Operations Record Book, 455 Squadron Archives Files and the personal records of RAAF aircrew. The list remains incomplete.

Appendix 6

German Naval Vessels

Torpedoboot - **Torpedo Boat.**

There were several classes of torpedo boats which were used for offensive sweeps and patrols, escort duties and mine-laying. They were between 1,132 tons and 2,587 tons and could make up to 37 knots. Torpedo boats could be armed with up to four 105 millimetre guns, four 37 millimetre guns, two twin-mounted 20 millimetre anti-aircraft guns, one quadruple-mounted 20 millimetre anti-aircraft gun and two sets of triple torpedo tubes. Germany possessed over thirty of these boats at one stage during the War, and this figure included eight Elbing class vessels that were called destroyers by the Royal Navy.

Schnellboot - **E-boat**

E-boats were fast motor torpedo boats displacing up to 95 tons and capable of making between 35 and 40 knots. Operations by E-boats were purely offensive, with torpedo or mine, and they operated mainly at night, endeavouring to reach their patrol areas and then return to base in darkness. They operated from two main bases, Ijmuiden on the Dutch coast and Cherbourg on the French coast. They normally operated in flotillas, splitting into twos or threes when they reached their patrol areas. They could be armed with one 37 millimetre gun and up to three 20 millimetre guns.

Raumboot - **R-boat**

These were small motor minesweepers, similar in appearance to E-boats but much slower, cruising at up to 15 knots and with a top speed of up to 25 knots. They were between 115 and 189 tons and had no torpedo tubes. They had been found in many of the North Sea and Channel ports and were mainly employed in the primary role of mine-sweeping and mine-laying. They were also used in escort duties and were sometimes seen off the Frisian Islands. They were typically armed with up to two 37 millimetre guns, three twin-mounted 20 millimetre and machine guns.

Minensuchboot - **M-Class Minesweepers**

M-Class minesweepers were employed in minesweeping, escort duties, mine-laying and anti-submarine work. They displaced between 600 and 888 tons, and could make between 17 and 18 knots. Although specifically designed and equipped for minesweeping, they frequently provided anti-aircraft escort for convoys. When escorting convoys they frequently flew balloons. They could be armed with up to two 105 millimetre guns, one 37 millimetre anti-aircraft gun, one quadruple-mounted 20 millimetre anti-aircraft gun, four twin mounted 20 millimetre anti-aircraft guns and four rocket dischargers.

Sperrbrecher - **Mine Clearance**

These were specially strengthened ex-merchant vessels between 1500 tons and 8,000 tons. Germany had some 50 to 60 of these vessels whose main function was minesweeping. They were heavily armed with 88 millimetre guns, 37 millimetre guns, 20 millimetre guns and machine guns. Their formidable anti-aircraft armament was for self defence. They were to be found mainly in the Baltic and the North Sea, keeping open the shipping channels. They could be identified by their dazzle painting, cut down masts and many gun positions.

Vorposterboote - Trawler-type Auxiliaries

A large number of the minor war vessels employed by Germany were converted trawlers. Their role could be minesweeping, anti-submarine patrol or convoy escort. They varied substantially in length and tonnage and to a certain extent in armament. A typical armament would be one 88 millimetre gun, two or three 20 millimetre guns and a number of machine guns. They were capable, in many cases, of up to 12 knots. Their obvious gun positions distinguished them from trawlers engaged in fishing. When employed on convoy escort they frequently flew balloons. Some made use of a parachute projectile and others had a form of flame-thrower at the mast head.

Armed Coasters

Armed coasters were usually coasters converted for cargo carrying and were armed with a 105 millimetre gun and up to six 37 millimetre anti-aircraft guns for self defence. They were recognized by a mast and derrick amidships.

Notes

Abbreviations:

ORB   Operations Record Book 455
Squadron RAAF, AWM 64
CCR   RAF Coastal Command Review
AA   Australian Archives Files (ACT
Repository)

1 455 Squadron The Hampden Years

1. Notes from a meeting to discuss the possibilities of a comprehensive rationalized Dominion training scheme, 22 September 1939, Air 2/3158, Public Record Office (PRO), London. This information is from an article *The 'surrender' of aircrew to Britain 1939-45* by Dr J. McCarthy, Australian War Memorial Journal, Issue 5 dated October 1984
2. J. McCarthy *A last call of Empire* Australian War Memorial, Canberra, 1988, p. 12
3. Sir R. Williams *These Are Facts* Australian War Memorial, Canberra, 1977, p. 245
4. J. McCarthy *The 'surrender' of air crew to Britain 1939-45*, p. 3
5. R. G. Menzies *Empire Air Force: Australia Plays Her Part* Melbourne, 1939 (broadcast 11 October 1939). From an article *The 'surrender' of aircrew to Britain 1939-45* by J. McCarthy
6. Of these, there were to be 3,100 pilots, 2,000 observers, and 3,300 gunners to be trained in Australia each year. 80 pilots, 42 air observers and 72 air gunners were to go to Canada every month for advanced training.
7. J. Herington *Air War Against Germany and Italy 1939-1943* Australian War Memorial, Canberra, 1954, p. 4
8. J. McCarthy *The 'surrender' of air crew*, p. 3
9. G. Long *To Benghazi* Australian War Memorial, Canberra, 1986, p. 59
10. W. Waldock interview
11. P. Ilbery interview
12. J. Lawson, Unpublished paper *455 Squadron RAAF* held in the Historical Section, Air Force Office, Russell Offices, Canberra

13. ibid., p. 6
14. G. Lindeman letter
15. Herington *Air War Against Germany and Italy* p. 116
16. T. Robertson *Channel Dash* The Quality Book Club, London 1958, p.35
17. AA1969/100/96, File 140/26/PI. Perrin's crew were F/O E.G. de T. Symons RAF (Wireless Air Gunner), P/O A.R. Abbot RNZAF (Navigator), and SGT R.E. Tomlinson (Gunner)
18. G. Lindeman letter
19. T. Robertson *Channel Dash* p.185
20. J. Herington *Air War Against Germany and Italy* p. 269
21. T. Robertson *Channel Dash* pp. 185 -186
22. G. Lindeman
23. ibid.
24. Bob Hay became Gibson's navigator for the Dambuster operation. Grant Lindeman has recalled that it was Hay who devised the means of accurately measuring the height of the bombers above the water, which was critical for the success of these raids.
25. Raeder, Grand Admiral *Struggle for the Sea* Translated from the German by Edward Fitzgerald, William Kimber, London, 1959, p. 157
26. J. Terraine *The Right of the Line* Hodder and Stoughton, London 1985, p. 407. Terraine refers to a report of the Admiralty/Air Ministry Committee on Anti-Submarine Warfare
27. J. Davenport, interview
28. ibid.
29. G. Lindeman letter
30. J. Davenport interview
31. ibid.
32. J. Lawson *The Story of 455 (R.A.A.F.) Squadron* Wilkie, Melbourne, p. 71
33. J. Davenport interview
34. J. Lawson *The Story of 455* p. 169
35. ibid., p. 173
36. Smart's crew were Sgt J.M.O. Harris (Wireless Operator/Air Gunner), Sgt T.G. Nicholls (Navigator), Sgt L.A. Biggin (Wireless

Operator/Air Gunner) and Cpl D. Nelson (Fitter). The bodies of the crew were recovered and buried in Stockholm.

37. J. Davenport, comment on draft
38. J. Davenport interview. This offensive patrol took place on 14 September 1942.
39. ibid.
40. P. Joubert *Birds and Fishes* Hutchinson, London, 1960, p. 133
41. CCR Vol 4 No 6 p. 13. 254 Squadron RAF, with a full complement of thirty-one Beaufighters, was armed with torpedoes. Both 143 and 236 Squadrons RAF, armed with cannon, formed the anti-flak section.
42. AA/1969/100/96, File 36/6/AIR
43. J. Slessor *The Central Blue* Cassell 1956, p. 547. Quoted in Terraine *The Right of the Line* p. 422
44. In February 1943 Wing Commander Lindeman was posted as the Chief Instructor at the Torpedo Training Unit at Turnberry in Ayreshire. He was awarded the Distinguished Flying Cross for leading 455 Squadron on the Russian operation. Later, he would be admitted to the Order of the British Empire for his work as Officer in Charge of Training at 19 Group (Coastal Command) during the crucial period leading up to the Allied invasion of Europe in 1944. He is one of the few Australian airmen to have been Mentioned in Dispatches three times.
45. D. Whishaw letter

2 First Beaufighter Operations

1. J. Davenport interview
2. L. Wiggins letter
3. P. Ilbery interview
4. J. McKnight interview
5. ORB 12 December 1943
6. ibid.
7. ORB 30 January 1944. The Airspeed Oxford Mk II was a twin-engine trainer
8. AA1969/100/96, File 136/PI
9. ibid.
10. Grand Admiral K. Doenitz *Memoirs* Greenhill Books, London, 1990, p. 407
11. S. Milson letter
12. J. Herington *Air War Against Germany and Italy 1939-1943* p. 397, and Citation of award of the Distinguished Flying Cross to Acting Squadron Leader C.G. Milson dated 20 March 1943
13. C.G. Milson letter to his father 21 May 1943, courtesy S. Milson
14. AWM 220. The vessel was the 944 ton merchant ship *Rabe*

15. ORB 13 April 1944
16. J. McKnight interview
17. D. Tunnicliffe *From Bunnies to Beaufighters* Alan Tunnicliffe, Christchurch, 1990, p.181
18. Marshal of the R.A.F. Sholto Douglas *Years of Command* Collins, London, 1966, pp. 262 - 268
19. CCR April 1944 p. 1
20. ORB 4 May 1944
21. J. Davenport interview
22. ibid.
23. AA1969/100/96, File 10/AIR
24. ORB 6 May 1944
25. The 1456 ton *Edouard Giess* was sunk
26. N. Smith letter
27. AA1969/100/96, File 11/2/Air SURFAT Report
28. AA1969/100/96 Recommendation for Award
29. CCR Vol III, No 5, May 1944, p.1
30. ibid. Vol III No 10, p.13
31. J. Davenport interview
32. AA1969/100/96
33. AA1969/100/96, File 121/PI

3 D-Day, 'the air must hold the ring'

1. CCR October 1944, p.12
2. K. Ward letter
3. N. Smith letter
4. I. Gordon interview
5. Recommendations for Honours and Awards John Charles Payne 11 June 1944 AA1969/100/96, File 107/PI
6. R. Andrew interview
7. Recommendations for Honours and Awards John Charles Payne 11 June 1944
8. E. Collaery letter to his wife, 12 June 1944
9. N. Turner letter
10. AWM 54 779/3/129
11. N. Turner letter
12. J. Ayliffe letter
13. D. Whishaw letter
14. Air Vice-Marshal A. Ellwood, from a lecture given to the United Services Institute 8 November 1944 and printed in the Coastal Command Review
15. CCR Vol III, Number 9 pp. 14 - 15
16. D. Whishaw letter
17. C. Bowyer *Beaufighter at War* Ian Allan Ltd, Surrey, 1976, p.122
18. C. Milson BBC broadcast recorded on 78 rpm record, courtesy S. Milson
19. BBC Broadcast *With the Australians in Britain* transcript courtesy S. Milson
20. ibid.
21. CCR June 1944 p.1

22. Recommendation for Immediate Award of Bar to Distinguished Flying Cross for Acting Squadron Leader C.G. Milson dated 16 June 1944

23. E. Watson diary, courtesy M. Watson

24. Before arriving at an operational unit, all aircrew had to pass through three or four schools and training units, and most took up to two years from enlistment to joining their operational squadrons. The courses included basic training, elementary flying training, service flying training and reconnaissance training. Aircrew continued to an advanced flying unit and then finished at an operational training unit.

25. E. Collaery letter 2 May 1944, courtesy B. Collaery

26. J. Cox letter

27. ibid.

28. J. Cox interview

29. Cox letter

30. A. Ibbotson interview

31. J. Cox letter

32. J. Cox interview

33. H. Pearson interview

34. ibid.

35. Recommendation for Immediate Award of Distinguished Flying Cross H. W. Pearson, 6 April 1945, AA1969/100/96, File 107/PI

36. ibid.

37. Letter to Mrs A.L. Collaery from Wing Commander J. Davenport, 30 June 1944

38. Australian Red Cross Society, London Advisory Committee letter to Mrs Collaery 9 August 1944

39. J. Cox comment on draft

4 Rockbeau Squadron

1. J. Cox comments on draft

2. J. Davenport interview

3. J. Davenport letter

4. I. Gordon interview

5. D. Whishaw letter

6. ibid.

7. ibid.

8. ibid.

9. ibid.

10. Air Ministry Bulletin 14593 dated 7 July 1944

11. W. Kimpton comments on draft

12. E. Watson diary courtesy M. Watson

13. D. Whishaw letter

14. ORB 10 July 1944

15. BBC Recording courtesy S. Milson

16. ORB 15 July 1944

17. E. Watson diary

18. CCR July 1944 p.1

19. BBC recording, courtesy S. Milson

20. ORB 21 July 1944

21. D. Whishaw letter

22. 455 Squadron Line Book AWM 64.

23. AA1969/100/96, File 146/PI

24. E. Watson diary, courtesy M. Watson

25. D. Whishaw letter

26. Coastal Command Review July 1994 p.1

27. Doenitz *Memoirs* p. 396

28. J. Davenport interview

29. ibid.

30. ibid.

31. ibid.

32. Doenitz *Memoirs* p. 396

33. ibid., p. 397

5 First and Last Light Strikes

1. C. Wilmot *The Struggle for Europe* Collins, London, 1952, p. 406

2. Coastal Command's Strike Wings were deployed in August 1944 as follows:

Langham	455 Squadron RAAF (Beaufighters)	
	489 Squadron RNZAF (Beaufighters)	
North-Coates	143 Squadron RAAF (Beaufighters)	
	254 Squadron RAF (Beaufighters)	
	236 Squadron RAF (Beaufighters)	
Strubby	144 Squadron RAF (Beaufighters)	
	404 Squadron RCAF (Beaufighters)	
Portreath (Cornwall)	235 Squadron RAF (Mosquitoes)	
	248 Squadron RAF (Mosquitoes)	

3. D. Tunnicliffe *From Bunnies to Beaufighters* p. 195

4. ibid.

5. J. Ayliffe letter

6. K. Ward letter

7. H. St. George Saunders *Royal Air Force 1939-1945* Her Majesty's Stationery Office, London, 1954, p. 133

8. N. Turner letter

9. ibid.

10. ibid.

11. A. Ibbotson interview

12. I. Gordon interview

13. Air Ministry Bulletin 15579 dated 10 September 1944

14. J. Ayliffe letter

15. Air Ministry Bulletin 15579

16. ibid.

17. I. Gordon interview

18. J. Ayliffe letter

19. I. Gordon letter

20. J. Cox interview

21. D. Whishaw letter

22. J. Ayliffe letter

23. C. Milson letter to his father 22 December

1944, courtesy S. Milson
24. J. Tucker, comment on draft
25. CCR September 1944 p. 1
26. J. Ayliffe letter
27. J. Tucker interview
28. D. Tunnicliffe *From Bunnies to Beaufighters* p. 201
29. J. Tucker letter
30. N. Turner letter
31. J. Cox letter

6 'Into German Harbours'

1. CCR September 1944, p.1
2. D. Whishaw letter
3. ORB 12 September 1944
4. Air Ministry Bulletin 15623 dated 13 September 1944
5. ibid.
6. D. Whishaw letter
7. ibid.
8. R. Nesbit *The Strike Wings* William Kimber, London, 1984, p. 180
9. From C. Wilmot *The Stuggle for Europe*, p. 548-550
10. D. Whishaw letter
11. B. Iggulden letter
12. ibid.
13. D. Whishaw letter
14. N. Smith interview
15. D. Whishaw letter
16. B. Iggulden letter
17. A. Clouston *The Dangerous Skies* Cassell, London 1954, p. 173
18. A. Clouston *The Dangerous Skies* p. 173
19. I. Gordon interview
20. N. Smith interview
21. D. Tunnicliffe *From Bunnies to Beaufighters* p. 203
22. I. Gordon interview
23. ibid.
24. D. Tunnicliffe *From Bunnies to Beaufighters* p. 204
25. I. Gordon interview. Since joining 455 Squadron in August 1943, Colin Cock had completed 43 operational sorties. Bob Lyneham had completed 22 operational sorties since joining the Squadron in April 1944.
26. A. Clouston *The Dangerous Skies* p. 175
27. J. Davenport comments on draft.
28. ibid.
29. CCR September 1944 p. 1
30. L. Wiggins letter
31. E. Watson diary, courtesy M. Watson
32. A. Ibbotson letter
33. ibid.
34. E. Watson diary

35. S.W. Roskill RN *The War at Sea* Vol 3 Part 2 HMSO, London p. 60
36. The strike squadrons operating from Banff in October 1944 were:
143 Squadron RAF (Mosquitoes) from North Coates
235 Squadron RAF (Mosquitoes) from Portreath
248 Squadron RAF (Mosquitoes) from Portreath
144 Squadron RAF (Beaufighters) from Strubby
404 Squadron RCAF (Beaufighters) from Strubby
37. J. McKnight interview
38. ibid.
39. ibid.

7 Dallachy

1. J. Davenport interview
2. N. Turner letter
3. P. Ilbery letter
4. No. 16 Group Coastal Command Tactical Memorandum No. 5/1944 dated 16 October 1944, AA 1969/100/96, File 33/4/Air
5. J. Ayliffe comment on draft
6. J. Cox letter
7. E. Watson diary
8. J. Cox letter
9. J. Berry letter
10. ibid.
11. J. Berry, interview
12. N. Turner letter
13. J. Ayliffe letter
14. J. Davenport comment on draft
15. Headquarters No. 18 Group. 18G/S,2201/18 dated 27 December 1944, AA1969/100/96, File 33/4/Air
16. Air Ministry S.61140/III/S.7. (a) (1) dated 17 November 1944, AA1969/100/96, File 33/4/Air
17. J. McKnight interview
18. P. Ilbery letter
19. J. Cox letter
20. E. Watson diary
21. A. Ibbotson interview
22. J. Cox letter
23. E. Watson diary
24. Fuhrer Conference on Naval Affairs, 11 December 1944, quoted in F. Hinsley *British Intelligence in the Second World War* Cambridge University Press, New York, 1984, p. 495
25. W. Herbert comment on draft
26. J. Ayliffe comment on draft
27. D. Whishaw letter

8 Fiords and Enemy Fighters

1. E. Watson diary, courtesy M. Watson
2. P. Ilbery interview

3. ibid.

4. P. Ilbery letter

5. Pilot Officer Dunshea's body was recovered from a beach at Haramso, with the body of an RAF pilot. Pilot Officer Winter's body was not recovered. Pilot Officer Dunshea is buried in Trondheim (Stavne) cemetery in Norway

6. AA1969/100/1

7. C.J.M. Goulter *A Forgotten Offensive, Royal Air Force Coastal Command's Anti-Shipping Campaign, 1940-45*, Doctoral Thesis, King's College, University of London, 1993, Chapter 8. Dr. Goulter draws on several official sources.

8. via J. Cox

9. J. Tucker interview

10. C. Milson letter to his father 31 January 1945. 'Woodge' was the family nickname for Colin Milson.

11. The *Narvik* class destroyer was designated Z-33, of 2603 tons. It was armed with five 105 millimetre guns, six 37 millimetre guns and eight 20 millimetre cannons.

12. J. MacDonald interview

13. H. Spink interview

14. J. Tucker letter

15. Warrant Officer Donald Mutimer's navigator was usually Flight Sergeant Ivor Jury, who was not available on 9 February 1945 so Blackshaw was flying in his place. Flight Sergeant John Blackshaw's pilot was usually Warrant Officer Bill Mitchell, but this day Mitchell was in London playing football. Jury and Mitchell then teamed up and both were lost when their aircraft was shot down on 8 March 1945. Flight Sergeant Blackshaw and Warrant Officer Mutimer are both buried in Haugesund (Rossebo) Churchyard in Norway.

16. H. Spink ibid.

17. J. Tucker letter

18. H. Spink ibid.

19. AA 1969/100/96, File 107/PI

20. A note written by Bob Macbeth in the fly leaf of a book he had bought, courtesy L. Macbeth

21. AA1969/100/96, File 140/112/PI

22. R. Macbeth letter to D. Whishaw, 19 February 1945

23. AWM 54 File 779/3/129 Part 15

24. R. McColl Diary, courtesy N. McColl

25. K. Ward letter

26. AA1969/100/96, File 140/114/PI

27. P. Ilbery letter

28. AA1969/100/96, File 140/115/PI

29. P Ilbery letter

30. ibid.

31. CCR March 1945, p. 3

32. AA 1969/100/96, File 140/116/PI

33. Warrant Officer George Longland and Flight Sergeant Errol Nayda are buried in Egersund Churchyard in Norway

9 Strikes in Procession

1. Lord Tedder *With Prejudice, The War Memoirs of Marshal of the Air Force* Cassell, London, 1956

2. A. Ibbotson letter

3. J. Ayliffe comment on draft

4. Air Ministry Bulletin No 18299 dated 10 April 1945.

5. ibid.

6. A. Ibbotson letter

7. AA1969/100/96, File 107/PI

8. N. Turner letter

9. D. Whishaw letter

10. J. Ayliffe comment on draft

11. I. Harrison letter

12. *Victory Roll, The RAAF at War* Australian War Memorial, 1945, p. 152

13. I. Murray letter

14. ibid.

15. P. Ilbery letter

16. *Victory Roll* p. 152

17. J. Cox letter

18. ibid.

19. N. Turner letter

20. Flying Officer Thomas Higgins and Warrant Officer Alan Mirow are buried in Trondheim (Stavne) Cemetery in Norway

21. I. Gordon interview

22. ibid.

23. ORB 23 April 1945

24. J. Cox letter

25. ibid.

26. J. Tucker interview

Epilogue

1. S. Douglas *Years of Command* Collins, London, 1966, p. 277

2. J. Ayliffe letter

3. J. Cox letter

4. 455 Squadron Detail by Wing Commander C.G. Milson, serial number 24, dated 25 May 1945

5. C.J.M. Goulter *A Forgotten Offensive* Chapter 9

6. I. Gordon interview

7. J. McKnight letter

8. P. Ilbery letter to Sheriff of Morayshire 9 January 1991

9. P. Ilbery letter

10. ibid.

Bibliography

Official Sources

Australian War Memorial
AWM 54
AWM 64
AWM 88
AWM 220

Australian Archives, Series AA1969/100/96 Files

Coastal Command Review

British Air Ministry Bulletins:
14593 dated 7 July 1944
15579 dated 10 September 1944
15623 dated 13 September 1944
18299 dated 10 April 1945.

Books and Monographs

Bowyer, C. *Beaufighter at War* Ian Allan Ltd, Surrey, 1976

Bryant , A. *Triumph in The West 1943-1946* Collins, London, 1959

Bullock, A. *Hitler* Odhams Press, London, 1952

Clouston, Air Commodore A.E. *The Dangerous Skies* Cassell, London, 1954

Doenitz , Grand Admiral K. *Memoirs* Greenhill Books, London, 1990

Douglas, Lord Sholto *Years of Command* Collins, London ,1966

Frischhauer, W. and Jackson, R. *The Navy's Here, The Altmark Affair*, Victor Gollancz, London, 1955

Herington, J. *Air War Against Germany and Italy 1939-1943* Australian War Memorial, Canberra, 1954

Herington, J. *Air Power over Europe 1944-1945* Australian War Memorial, Canberra, 1963

Hinsley, F. et al *British Intelligence in the Second World War* Cambridge University Press, New York, 1984

Horner, D. *High Command* Allen and Unwin, Sydney, 1982

Joubert, Air Chief Marshal Sir P. *Birds and Fishes* Hutchinson, London, 1960

Lawson, J.H.W. *The Story of 455 (R.A.A.F.) Squadron* Wilke, Melbourne, 1951

Long, G. *To Benghazi* Australian War Memorial, Canberra, 1986

McCarthy, J. *A Last Call of Empire* Australian War Memorial, Canberra, 1988

Nesbit, R. C. *The Strike Wings* William Kimber, London, 1984

Raeder, Grand Admiral Struggle for the Sea Translated from the German by Edward Fitzgerald, William Kimber, London ,1959

Richards, D. and Saunders H. St George Royal Air Force 1939-1945, Vol. I, HMSO, London, 1953

Richards, D. and Saunders H. St George Royal Air Force 1939-1945, Vol. II, HMSO, London, 1954

Richards, D. and Saunders H. St George *Royal Air Force 1939-1945*, Vol. III, HMSO, London, 1954

Robertson, T. *Channel Dash* The Quality Book Club, London, 1958

Roskill, Captain S.W. RN *The War at Sea*, Vol 3 Part 2 HMSO, London, 1956

Roskill, Captain S.W. RN *The Navy at War 1939-1945* Collins, London, 1960

Saunders, H. St. George *Royal Air Force 1939-1945* Her Majesty's Stationery Office, London, 1954

Saunders, H. St. George *The Green Beret* Michael Joseph Ltd, London 1949

Slessor, J. *The Central Blue* Cassell, 1956

Tedder, Lord *With Prejudice, The War Memoirs of Marshal of the Air Force*, Cassell, London, 1956

Terraine, J. *The Right of the Line* Hodder and Stoughton, London, 1985

Terraine, J. *Business in Great Waters* Leo Cooper Ltd, London, 1989

Tunnicliffe, D.M. From *Bunnies to Beaufighters* Alan Tunnicliffe, Christchurch, 1990

Thompson, Wing Commander H.L. *New Zealanders with the Royal Air Force Volume II* European Theatre War History Branch Department of Internal Affairs New Zealand, Wellington, 1956

Victory Roll, The RAAF at War Australian War Memorial, Canberra, 1945.

Williams, Sir R. *These Are Facts* Australian War Memorial, Canberra, 1977

Wilmot, C. *The Struggle for Europe* Collins, London, 1952

Articles

McCarthy, J. *The 'surrender' of aircrew to Britain 1939-45*, Australian War Memorial Journal, Issue 5 dated October 1984

Unpublished Papers

Goulter, C.J.M. *A Forgotten Offensive, Royal Air Force Coastal Command's Anti-Shipping Campaign, 1940-45*, Doctoral Thesis, Kings College, University of London, 1993.

Lawson, J.H.W. *455 Squadron RAAF* Unpublished paper held in the Historical Section at Air Force Office, Russell Offices, Canberra

BBC Broadcast *With the Australians in Britain* Transcript courtesy S. Milson

Diaries

E.W. Watson (courtesy M. Watson)
R.C. McColl (courtesy N. McColl)

Interviews

R.J. Andrew, J.C. Ayliffe, J. Berry, J.G. Cox, J.N. Davenport, I.M. Gordon, W.G. Herbert, A.W. Ibbotson, P.L. Ilbery, J.A. McKnight, J. MacDonald, H.W. Pearson, N.R. Smith, H.R. Spink, J.W. Tucker, N.P. Turner, W.F. Waldock, D. Whishaw, W. Bremner, R.W. Kimpton, F. Macintyre, W.F. Waldock, A.L. Wiggins.

Letters

J.C. Ayliffe, J. Berry, J.G. Cox, J.N. Davenport, I.M. Gordon, I. Harrison, A. W. Ibbotson, P.L. Ilbery, W.H.H. Iggulden, R.E. Jones, G.M. Lindeman, R.D. Macbeth (to D. Whishaw), J.A. McKnight, C.G. Milson (to his father), S. Milson, I Murray, N.R. Smith, N.P. Turner,
J.W. Tucker, K. Ward, D. Whishaw, A.L. Wiggins, I.A. Murray, E.F.Collaery (to his wife).

Recordings

BBC Broadcast recording on 78 rpm disk of S/Ldr C.G. Milson and W/Cdr J.N. Davenport, July 1944, courtesy S. Milson

Index

www.ingramcontent.com/pod-product-compliance
Lightning Source LLC
Chambersburg PA
CBHW050458110426
42742CB00018B/3297